BODYSPACE

BodySpace

BodySpace b........ **DATE DUE** ....rs writing on gender and sexuality today. Together they explore the role of space and place in the performance of gender and sexuality.

The book takes a broad perspective on feminism as a theoretical critique, and aims to ground notions of citizenship, work, violence, 'race' and disability in their geographical contexts.

The book explores the idea of knowledge as embodied, engendered and embedded in place and space. Gender and sexuality are explored – and destabilized – through the methodological and conceptual lenses of cartography, fieldwork, resistance, transgression and the divisions between local/global and public/private space.

Contributors: Linda Martín Alcoff, Kay Anderson, Vera Chouinard, Nancy Duncan, J.K. Gibson-Graham, Ali Grant, Kathleen M. Kirby, Audrey Kobayashi, Doreen Massey, Linda McDowell, Wayne D. Myslik, Heidi Nast, Gillian Rose, Joanne P. Sharp, Matthew Sparke, Gill Valentine.

Nancy Duncan is an Affiliated Lecturer in the Department of Geography at Cambridge University.

BODYSPACE

destabilizing geographies of gender and sexuality

Edited by NANCY DUNCAN

LONDON AND NEW YORK

First published 1996
by Routledge
11 New Fetter Lane, London EC4P 4EE

Simultaneously published in the USA and Canada
by Routledge
29 West 35th Street, New York, NY 10001

Typeset in Photina by Keystroke, Jacaranda Lodge, Wolverhampton
Printed and bound in Great Britain by Clays Ltd, St. Ives PLC

British Library Cataloguing in Publication Data
A catalogue record for this book is available from the British Library

Library of Congress Cataloguing in Publication Data
BodySpace: destabilizing geographies of gender and sexuality/edited by Nancy
Duncan.
p.   cm.
Includes bibliographical references and index.
1. Feminist theory. 2. Feminist geography. 3. Masculinity (Psychology)
4. Sex role. I. Duncan, Nancy.
HQ1190.b63 1996
305.42′01–dc20   96–589

ISBN 0–415–14441–8 (hbk)
0–415–14442–6 (pbk)

CONTENTS

CONTRIBUTORS

Linda Martín Alcoff teaches Philosophy and Women's Studies at Syracuse University, USA. She is the co-editor, with Elizabeth Potter, of *Feminist Epistemologies* (Routledge 1993) and the author of *Real Knowing: New Versions of Coherence Epistemology* (Cornell University Press 1996).

Kay Anderson is Senior Lecturer at University College, University of New South Wales, Australia, where she teaches Cultural and other Human Geography courses. Since publishing work on race relations in Canada and Australia, she has examined women's political activism and is currently developing her research on animals and zoos in a study of nature's domestication in Australia.

Vera Chouinard is Associate Professor of Geography at McMaster University, Canada. Her research areas include: women, law and the state, cooperative housing policies, state regulation of community-based legal services, and disabled women's struggles. Her recent publications include: 'Geography, law and legal studies: which ways ahead?' *Progress in Human Geography* (1994).

Nancy Duncan is an Affiliated Lecturer in the Department of Geography at Cambridge University, UK. Her research interests are in political and cultural geography. She is co-author with James Duncan of *Suburban Pretexts: Deconstructing History, Localism, and Nature in a Westchester County Town* (forthcoming, Johns Hopkins University Press).

J.K. Gibson-Graham is the pen-name of **Katherine Gibson** and **Julie Graham**, industrial geographers who work on conceptions of capitalist society and their implications for gender and class politics. Gibson teaches in the Department of Geography and Environmental Science at Monash University, Melbourne, Australia, and Graham teaches in the Department of Geosciences at the University of Massachusetts-Amherst, USA. After working together for 15 years they adopted this joint writing persona in 1992, in a playful attempt to forge new individual and collective subjectivities within their collaborative relationship. They recently published *The End of Capitalism (As We Knew It)* (Blackwell 1996).

Ali Grant is completing a PhD at McMaster University, Canada, on 'Geographies of oppression and resistance: contesting the heterosexual regime'.

Kathleen M. Kirby has taught English and Women's Studies at the University of Wisconsin-Milwaukee, Syracuse University and the University of New Hampshire, USA. She is author of *Indifferent Boundaries: Spatial Concepts of Human Subjectivity* (Guilford Press 1995).

Audrey Kobayashi is Professor of Geography and Director of the Institute of Women's Studies at Queen's University, Kingston, Canada. Her research and publications are in the field of gender and racism, human rights and immigration.

Doreen Massey is Professor of Geography at the Open University, UK. Her most recent books include *Space, Place, and Gender* (Polity 1994), *Spatial Divisions of Labour*, (2nd edn, Macmillan 1995), *High-Tech Fantasies* (with P. Quintas and D. Wield, Routledge 1992), *Geographical Worlds* (ed. with J. Allen, Oxford and Open University Presses 1995), *A Place in the World? Places, Cultures and Globalization* (ed. with P. Jess, Oxford and Open University Presses 1995).

Linda McDowell teaches Urban and Social Geography in the Department of Geography at Cambridge University, UK. Her main interests are in the changing nature of work and in feminist theory. She is co-editor of *Defining Women* (with R. Pringle, Polity 1992) and author of a forthcoming book on merchant banking, *Capital Culture*. She is also co-editing a forthcoming book entitled *Space–Gender– Knowledge* with Joanne Sharp.

Wayne D. Myslik is a PhD candidate in the Department of Geography at Syracuse University, USA. His research interests are in Social and Cultural Geography with particular reference to race and sexuality.

Heidi J. Nast teaches in the International Studies Department at DePaul University, Chicago, USA. She has published in the areas of sexuality and the state and feminist cultural geography. She has worked in both the United States and West Africa.

Gillian Rose teaches feminist and cultural geographies at the University of Edinburgh, UK, and her research interests concern the politics of producing geographical knowledges. She is author of *Feminism and Geography* (Polity 1993) and, with Alison Blunt, co-editor of *Writing, Women and Space: Colonial and Postcolonial Geographies* (Longman 1994).

Joanne P. Sharp is a Lecturer in Geography at the University of Glasgow, UK. Her primary research interests are in political and cultural geography. She is co-editing the forthcoming collection *Space–Gender–Knowledge* with Linda McDowell.

Matthew Sparke is an Assistant Professor in Geography and International Studies at the University of Washington, Seattle, USA. His intellectual interests concern the intersection of social and cultural dynamics of identity formation with the economic and political geographies of late twentieth-century capitalism. His work is informed by feminist, Marxist and post-foundationalist theory, and he is committed to research that examines the practical negotiation of social change in Canadian and US communities.

Gill Valentine is Lecturer in Geography at the University of Sheffield, UK, where she teaches courses on society and space and qualitative methods. Her research interests include: geographies of sexuality; children and youth culture; and food and foodscapes. She is co-editor with David Bell of *Mapping Desire* (Routledge 1995).

ACKNOWLEDGEMENTS

I would like to thank all of the contributors to this volume for their thoughtful essays. I am grateful to the Department of Geography and the Maxwell School of Citizenship at Syracuse University for sponsoring the symposium entitled 'Place, Space and Gender' out of which this volume evolved and to David Robinson who originally suggested the idea of holding such a symposium. I would also like to thank Jim Duncan for encouraging me to work on this project even though it took time away from another book I was supposed to be working on with him and for involving me in his seminar on Feminisms in Geography which laid the ground work for the symposium. Joanne Sharp and two reviewers of the manuscript offered many valuable comments on an earlier draft which contributed to an improved final product. I also wish to thank Tristan Palmer, my editor at Routledge, for his encouragement and support.

I gratefully acknowledge permission to reproduce copyright material from *Antipode* and *Aperture*.

INTRODUCTION

(Re)placings

Nancy Duncan

Feminists are presently exploring the far-reaching implications of a new epistemological viewpoint based on the idea of knowledge as embodied, engendered and embedded in the material context of place and space. This requires not amendments or additions to allegedly universal (but in actuality masculinist and often Eurocentric) discourse, nor a reversal, but a 'strategic transformation' (Alcoff, this volume). The contributors to this volume wish to help push this project forward through a reconsideration and re-politicization of such geographical concepts as space, place, the local and the global, sites of resistance, cartography, fieldwork, the transgression of boundaries, and the public/private division of space. Thus, as the title suggests, *BodySpace* intends to 'place' gender and sexuality (in both corporeal and discursive terms) squarely on the academic agenda by emphasizing place, space and other geographic concepts that are useful in contextualizing and situating social relations.

Adopting a broad perspective on feminism as a theoretical critique, and a political perspective which aims to destabilize and transform gender relations and hierarchies, the contributors to this volume recognize that daunting as this project may be in itself, it cannot be carried forth in isolation from other equally compelling progressive projects. As Alcoff and Potter (1993: 4) have stated:

> We find a strong consensus among feminists today that both the term and the project of feminism itself must be more inclusive than a focus on gender alone permits. If feminism is to liberate women, it must address virtually all forms of domination because women fill the ranks of every category of oppressed people. Indeed the ontological status of woman and even of women has shifted for academic feminists in light of influential arguments showing that women, per se, do not exist.

Questions of subjectivity, sexuality, masculinity, 'race', class, disabilities and nationalism thus figure prominently in the chapters that follow.

The first section of the book raises broad epistemological questions that are then explored in the next few chapters which address such issues as citizenship, work, disabilities and sexual minorities. Many of the same epistemological issues raised in the first section are then taken up again with regard to researching or 'doing' geography in the final section on engendering race research, masculinity and fieldwork, and postmodern feminist field research.

Universal categories of reason and knowledge as well as history and power are in fact reflections of gendered practices marked not only by gender but also by differences within gender. This gendered dualism of mind and body has spatial corollaries in other dualisms such as interiority/exteriority and public/private distinction. This latter distinction in turn depends upon other gendered dichotomies such as immanence/transcendence. Thus, while the public sphere has been seen as the sphere of universal reason and transcendence of the disembodied, disinterested Cartesian observer, in fact this model observer can be shown to be (implicitly) a white, bourgeois able-bodied male, and, in fact, as will be shown in several of the essays below, a heterosexual male.

Many feminists today believe that the goals of earlier generations of feminists who sought greater access for themselves and other 'Others' to this elite, male-dominated public sphere need to be reformulated. We now recognize that there can be no pure public spaces in which the liberal ideals of equality, impartiality and universality are achieved. In liberal theory the necessary homogeneity could only obtain if subjectivity (which is seen as stemming from particularities of bodily difference) were excluded and objectivity thus attained. Those marked by differences deriving from their sex, skin colour, old age, sexuality, physical incapabilities or other variations from the posited 'norm', do not qualify for full participation in the liberal democratic model. The materiality of our bodies is seen to exclude us from participating in an ideal of reason which 'knows no sex', no embodied differences. As Donna Haraway (1991: 184) puts it:

> the imagined 'we' are the embodied others, who are not allowed not to have a body, a finite point of view, and so an inevitably disqualifying and polluting bias in any discussion of consequence outside our own little circles . . .

Iris Marion Young argues that the ideals of liberal political theory such as formal equality and universal rationality and impartiality express 'what Theodor Adorno calls a logic of identity that denies and represses difference'. This repression, she argues, relies on: 'an opposition between public and private dimensions of human life, which corresponds to an opposition between reason, on the one hand, and the body, affectivity, and desire on the other' (1987: 63). The ideal of universality and impartiality, as Young and others have argued, is based upon a model of an individual abstracted from any real context – an individual as Young puts it, who has 'no particular history, is a member of no communities, has no body' (1987: 60). Those who come closest to the abstracted ideal human are able to dominate the public sphere. In fact even those who are included must separate their allegedly sexually neutral minds from their bodily needs and desires, these normally to be serviced in private spaces by others who are consequently less autonomous (Brown 1995: 156–65).

The reason why women tend to be among the excluded and why women who have managed to be admitted to active participation in the public sphere have usually done so according to implicitly male rules, has to do with the mind/body dualism. This dualism and its spatial counterpart – the public/private division – can be shown to be an important structuring principle upon which characteristics commonly associated with masculinity and femininity are arrayed. Thus public reason is associated with a supposedly neutral observer. The problem with this model, as a number of the essays below argue, is that this neutral standpoint is a fiction. Furthermore, politically and ethically it may be suspect even as an ideal that societies might strive to closely approximate. It is based on a false assumption of homogeneity that erases very real differentials in power and thus tends to conserve structured inequalities (see Young 1987). It is also based on a fiction of coherence and spatial and social centredness that erases cultural alterity, hybridity, marginality, distance and deterritorializing global processes.

The epistemological frameworks of various social sciences need to be transformed to accommodate an inclusion of women not as a special case deviating from the norm, but as one of many different groups in an open and heterogeneous universe. With respect to the collection of information and the generation of knowledge claims, Elizabeth Gross (1987: 191) has argued that:

> it became increasingly clear that it was not possible to simply include women in those theories where they had previously been excluded, for this exclusion forms a fundamental structuring principle and key presumption of patriarchal discourses. Many patriarchal discourses were incapable of being broadened or extended to include women without major upheavals and transformations. There was no space within the confines of these discourses to accommodate women's inclusion and equal participation.[1]

Many feminists favour the concept of situated knowledge as a substitute for decontextualized, disembodied, ungendered, 'objective' knowledge. Since the contributors to this volume are either geographers, or influenced by the geographical literature, they find this concept especially compelling for its inherently geographical claims. They welcome the current academic climate in which there is a widespread movement to recontextualize theory and a renewed interest in empirical work that pays close attention to geographical and historical specificity. In this regard, however, Gillian Rose (this volume) provides an important cautionary statement about geographers' insistence on grounding spatial metaphors and calls for a feminist deconstruction of the controversial distinction between metaphorical or discursive and so-called 'real' space. She argues that: 'The "real", "experienced", "concrete", "social", "actual", "geographical" are not the same things, and nor is any one consistently aligned with a non-real opposite' (Rose, this volume).

Situated knowledge or standpoint theory, terms often associated with the work of Donna Haraway (1991) and Sandra Harding (1986, 1991, 1993), can be seen as offering a considerable challenge to the universalistic pretensions of mainstream epistemology. Situated knowledge is knowledge that is 'located' by researchers who self-critically attend to the cultural, geographical and historical specificity of the conditions of production of those knowledge claims.

This view has been defended by Sandra Harding who argues that greater objectivity is achieved when one positions oneself acknowledging the impossibility of impartiality and disinterestedness. If one is a member of a dominant group, one can empathetically 'start off thought' from the lives of marginalized people (see Harding 1993). Harding contrasts this epistemological perspective with four other grounds for knowledge, 'God-trick, ethnocentrism, relativism, and the unique abilities of the oppressed to produce knowledge' (1993: 57). While social scientific knowledge is necessarily contextual, it conventionally presents itself as a 'view from nowhere'. Harding argues that conventional science may succeed in erasing the biases of particular scientists, but it sets up no procedures for removing the biases shared by the community of scientists or for seeking out observers whose standpoint and interests differ significantly. Harding (1993: 57) states, 'thus culturewide assumptions . . . are transported into the results of research'.

Contrary to what is commonly thought, relativism, ethnocentrism and an assumption that the oppressed occupy an unassailable moral high ground (that they alone can produce unbiased objective knowledge) are not the only alternatives to this conventional 'God-trick' (see Harding 1993). In other words, contra relativism, there is no radical alterity; although cultural and gender differences may not be assimilable, they are not incommensurable either.

As David Harvey (1993) suggests we need not assume that 'none of us can throw off even some of the shackles of personal history or internalize what the condition of being "the other" is all about'. In fact the attitude that we cannot be empathetic, he argues, would lead to an essentialist, exclusionary politics. 'Starting off thought' from the point of view of the oppressed or marginalized does not trade on the notion of authenticity; it is more like doing the hard work of 'earning the right' (Spivak 1988) and the ability to speak about others. Hegel's master/slave dialectic assumes that the oppressed can better understand the nature of power relations (in part) because they have less stake in apologizing for the status quo than do the oppressors. However, as Hartsock (1983: 284), Harding (1993: 62–9) and Harvey (1993: 108–9) suggest, one need not be a member of a subaltern group to create a subversive science from the perspective of that group.

If social researchers are to situate their knowledge claims in a socially progressive rather than solipsistic or self-authorizing way, consideration of the social, spatial, political and historical situation, and limitations of one's knowledge claims must become an integral part of the research process. Whether experience or empathy (usually both) are involved, difference is articulated and situated producing explicitly rather than covertly biased research. This way both intellectual rigour and progressive goals are more likely to be accomplished.

Social relations, including, importantly, gender relations, are constructed and negotiated spatially and are embedded in the spatial organization of places. One of the primary aims of this book is to address questions of space and place in order to show how the implicitly universalized claims of social scientists can be specified and qualified by paying close attention to differences. These differences are not only gender differences, but cultural and historical differences within gender, including sexuality which itself appears to be as variable historically and culturally as does gender. These multiple differences can in part at least be explained by looking at the construction of identities in diverse places – interconnected sets of places ranging in scale from the local to the global.

As Judith Butler (1992: 1–16) and many other feminists have argued, gender is a permanently contested concept. 'Woman' and 'man' are unstable categories which are only loosely and contingently related to sexuality. Heterosexuality is being increasingly denaturalized as a result of the efforts of gays, lesbians and bisexuals, as well as by those who reject hetero- or homo-sexual orientation as the basis for a stable identity or identity politics.

According to Butler gender is a cultural performance – the effect of a set of contested power relations based in 'such defining institutions' as 'phallogo-centrism and compulsory heterosexuality' (Butler 1990: viii). Butler proposes a performative theory of gender that disrupts the categories and correlations between bodies, sex, gender and sexuality. The authors of the essays in this book show in various ways how such performances and contests around power relations take place in lived space. Spatiality constrains, enables and is constituted by forces that both stabilize dominant relations of gender and sexuality and that unsettle the relations between them.

In her essay for this volume Linda Alcoff states that reason has been defined in opposition to what have traditionally been considered feminine characteris-tics and that the mind–body dualism is a central feature of the masculinist formulation of reason. She outlines four major premises of a new feminist theory of reason: that the mind and body are not separable, that mind has therefore never been separate from the body, that our dominant ideals of reason are reflections of embodied ways of being, and that we therefore have to rethink the many assumptions which pervade social science that are based on the mind–body dualism. She points out that because bodies are sexually specific and sociocultural, all knowledge claims which have been premised on the ideal of a disembodied, gender-neutral universal human being must be radically rethought. Alcoff then proceeds to provide examples of how the epistemological frameworks of various social sciences might be transformed to accommodate an inclusion of women not as a special case deviating from the norm, but as one of many groups in an open and heterogeneous universe.

Linda McDowell's essay follows with an argument for situating feminist theory through the use of spatial concepts that refer to space, place and *dis*placement: time–space distancing, fragmentation and intermixing such as migration, discontinuous multinational realities, a third space and interstitial 'in-between' places[2] where 'the West and the rest' often share the same spaces contesting coherent identities as a result of colonial, post-colonial and increasingly globalized economies. McDowell proposes the concepts of a 'global localism' and 'a geometry of multiple difference' which, as she argues, recognizes the unevenness of the disruptive impact of interconnected global capitalism on particular localities, knowledges, and place-based identities.

Kathleen Kirby focuses on cartography, placing what she names the Cartesian 'mapping' subject into historical, cultural and gendered perspective. Mapping space as a signifier of control is contrasted with bodily immersion and the sort of hybrid spaces that McDowell discusses. Kirby argues that Enlightenment rationalism developed the notion of a dominating, self-contained, masculine ego who maps out the world around himself and, in the process of charting paths and drawing boundaries, tends to exclude or marginalize non-dominant others who are more immersed in their environment and more aware of their embodiment. Kirby's argument adds a new dimension to Jameson's characterization of postmodernity and the cognitive mapping he sees as required to negotiate one's way in the postmodern world. The contemporary environment appears to him uncomfortably fluid and increasingly difficult to keep at arm's length. She points out that the newness of this uncomfortable feeling of immersion and embodiment in place and space may be related to Jameson's gender as well as his bourgeois whiteness. As she says 'women, the working class, and people of the Third World create a material environment for Western men, so they are able to expel it from their consciousness'. Women and others who are less able to control their environment, but are more responsible for its production and reproduction, have long been used to the feeling of overwhelming immersion. Such a breakdown of masculinist mapping power characteristic of the postmodern world (if Jameson's experience is in any way generalizable) may then lead to what Kirby (this volume) foresees as the promise of the postmodern: 'its tendency towards flux and revision; its porousness of division; the fluidity of its boundaries. The inclusive transformations we imagine might require eradicating, radically, the ordering lines of our culture, and our selves'.

Gillian Rose explores what she calls 'spatialized performances' that produce gender and argues that other performances can destabilize gender. Her paper is an engagement with Irigaray as 'a theorist of the spatial' in which she sets up a dialogue with Irigaray about the relation of space and gendered identity and then performs it for us, her readers. She discusses the creation of new forms of space based on 'all enveloping' notions of 'between' and 'around' – spaces that are supportive and enabling in contrast with notions of 'distance and separation' that affirm an individualistic 'master identity'. Rose's notion of all-enveloping space resonates with Kirby's postmodern space, as does her view that one of the important tasks of feminism is to subvert more bounded and proprietary space of the master subject.

Heidi Nast and Audrey Kobayashi discuss Jonathan Crary's (1993) theory of the history of vision with an eye to engendering this history. They show that the universalized observer that Crary takes to be a subject position produced by nineteenth-century discourse around vision is in fact much more specific than that. Following Crary and drawing on the work of Caroline Merchant (1983) and other feminists such as Susan Bordo (1987) they embody this subject, insisting that this body is a sexed body. They argue that corporealized 'differences' are integral rather than incidental to the history of vision.

In the second section the project of embodying, engendering and embedding is continued in six chapters which explore further many of the same concepts discussed in the first section. Each of these chapters focuses on somewhat more specific issues: citizenship, work, domestic violence, marginalized sexual identities and disabilities.

Joanne Sharp's essay considers the articulation of gender with national identity and citizenship. Sharp argues that the nation and citizenship are produced and reproduced through the repetition of symbols that reinforce and naturalize national identity. She critically re-examines and engenders Benedict Anderson's classic analysis of nationalism as an imagined community. By deconstructing the narrative of bonding between individuals and the nation, Sharp shows that the rhetoric of this bonding reveals that it is implicitly differentiated by gender.

Focusing particular attention on gender and nationalism in Eastern Europe, Sharp looks at the spatiality of gendered identities at various scales and in institutional sites from private domestic spaces to the space of the nation-state and beyond into the realm of international relations. She argues, following Foucault and in agreement with a point that Alcoff makes in her essay, that political relations of power and resistance operate at all scales and certainly not just in the public sphere. Such a perspective on power as dispersed, inhering in informal as well as formal political practice, is integral to the transformative feminist project as it is conceived in this book.

As Doreen Massey points out, one of the most important themes of contemporary feminism is the critique of the dualistic thinking which characterizes much of modern Western society (on this see Lloyd 1984). Massey says that many of these dualisms structure gender relations, even those dualisms that at first sight may seem to have little to do with gender. The most valued side of these dualisms is usually characterized as masculine and the less valued – feminine.

In her chapter, Massey examines the way two particular dualisms structure masculinity in particular and gender relations more generally, in the context of high-tech industry. The first is the dualism of Reason and non-Reason. This is closely related to another dualism – that of transcendence and immanence. Transcendence refers to progress, the making of history, scientific breakthroughs and the like. Transcendence is most often associated with men, especially privileged and talented men. Immanence is associated with embodiment, reproduction, servicing others and what Massey refers to as 'static living-in-the-present'. Although immanence is obviously the less socially valorized pole of the dualism, one can see the similarity between it and the experience of being enveloped by one's environment which Kirby and Rose revalue from a feminist perspective.

In my own chapter I make a call, similar to Kirby's, for a spatial revolution or a deterritorialization that would undermine boundaries between public from private spaces, thus (re)politicizing both private and public spheres and corresponding spaces. I first outline the always already unstable distinction between public and private spaces and the relation of these to public and private spheres. The public and the private constitutes one of the gendered dualisms which the feminist project outlined in this book seeks to destabilize. I then attempt to show the effects of this spatially structuring binary that is employed to exclude, control, confine and suppress gender and sexual difference preserving traditional patriarchal and heterosexist power structures. I explore the implications of spatially constituted gender and sexuality and show how the spatial practices of marginalized groups such as abused women and sexual minorities (lesbians, gays and sex workers) help to undermine the deeply rooted public/private spatial distinction. I then argue the case for a spatial revolution – a deterritorialization that would open and revitalize the public sphere making it more accessible to oppositional social movements and at the same time expose the contradictions of privacy.

Gill Valentine seeks to further politicize the question of heterosexist public spaces through her discussion of sexual dissidents: lesbians who perform acts of subversion and resistance that aim to renegotiate the (hetero)sexuality of everyday streets. She emphasizes the political aspects of the performance of lesbian sexuality that disrupts binary gender identities which naturalize heterosexuality. She shows how compulsory heterosexuality constructs space and how lesbians can destabilize and renegotiate the way spaces are used.

Wayne Myslik concentrates his attention on spatial strategies to cope with violence and the fear of violence against gays in public places. He indicates that gay men tend to monitor their behaviour in public in order to conform to the heterosexist illusion that sexuality is found only in private spaces. As Valentine discusses (this volume and 1993) the heterosexuality of public spaces that is alienating for gays tends to be naturalized almost to the point of invisibility for heterosexuals. Myslik shows that much of the violence against gays is perpetrated by 'average boys exhibiting typical behaviour'. Gay-bashing may not be the norm, but there is some evidence to show that most of those who commit this type of crime show no other criminal tendencies. Furthermore, he argues that they have been socialized within American society to express their heterosexual masculinity through dominance and aggression. Their attacks on gays both confirm and perform their heterosexuality for their peers. The attacks also play a role in policing and disciplining public behaviour, to force it to conform to the societal illusion of disembodied, desexualized public spheres and spaces.

Myslik goes on to discuss an interesting counter-intuitive finding of his own research in the gay community of Dupont Circle, Washington, DC. Dupont Circle can be described as a queer space because of the presence of gays who define it as a safe, non-alienating place where they do not feel the need to closely monitor their behaviour so as to avoid offending heterosexual norms. He found that such queer spaces are considered safe by gays even though they are specifically targeted areas of heterosexist violence. Apparently the sense of community spirit, emotional support and the vision of Dupont Circle as a site of cultural resistance against heterosexist norms of public behaviour overcomes fears of harassment and violence.

The chapter by Vera Chouinard and Ali Grant demonstrates how an 'unmarked' perspective based on assumptions about the human norms of able-bodiedness and heterosexuality produces spatial arrangements that unnecessarily handicap those who do not fit the ideal type. People who are marked by physical 'disabilties' have spatial and built form needs that do not match the norm, that is the implicitly privileged, young, healthy, male norm. All human beings have limited mobility and physical abilities. So-called disabilities and 'special' environmental needs, however, are defined and measured by the extent that these vary from this norm. Our public spaces potentially could be accessible to a much broader range of needs than they are at present. Instead the tendency is to 'blame the victim' by assuming that the disjuncture between the built environment and the environmental needs of the individual is due to the individual's inability to meet the 'normal' standard. If social and political values were more enlightened it would be assumed that the problem lay in spatial frameworks that neglect to fulfil the needs not only of minorities such as the physically 'disabled', and the elderly, but the less valued members of society including all types of women.

They go on to show that this unmarked perspective equally normalizes the hegemonic heterosexuality of most environments. Such normalization makes lesbians either invisible or – if they choose to signal their sexuality – they must be constantly under the exhausting pressure and responsibility of political struggle over the definition of space. As long as lesbians and the disabled remain invisible, radical geographical explanations of oppression will remain unnecessarily homogeneous and insensitive to differences among those who are marginalized and oppressed.

In the last section three authors provide reflections on research that does not pretend to be conducted from a disembodied, universalistic point of view. Kay Anderson discusses engendering 'race' research and critically reflects on her own previous research in Vancouver's Chinatown (1991) in which she feels that she had insufficiently crosscut the issues of gender with those of 'race' that had been the primary focus of her attention. Her brave reappraisal of her own book is an instance of critical self-reflexivity all too rare in academia. She provides an example of the difficult task of conceptualizing the intersection between gendered and 'racialized' positionings that necessarily disrupts modernist notions of 'racial' self–Other categories undifferentiated by class and gender. Like many of the other authors in this volume, Anderson emphasizes the multiple and fluid identities of the subjects of her study.

Matthew Sparke critically reflects upon the heroic masculinity of the spatial practice of field research. He investigates the implications of military analogies in masculinist conceptualizations of the field in geographical research, its 'gaze', and appropriative arrogance. He self-critically positions himself as a male feminist researcher studying the experiences of workers in an industry that predominantly hires women. Sparke wishes to (re)place conventional (read masculinist) concepts of fieldwork with a more dynamic and politically progressive notion of fieldworking in what he follows Kim England (1994) in calling 'the world between ourselves and the researched'.

J.K. Gibson-Graham's chapter focuses on replacing modernist research methods with epistemologically challenging (and politically charged) postmodernist approaches that deconstruct essentialized notions of 'woman'. Necessarily multiple, fragmented and decentred subjectivities posited by such a deconstructed concept of woman are difficult to represent; as she puts it, 'the objects of our research dissolve before our eyes'. The challenge is to destabilize the overly disciplined subject positions which currently inform prevailing attitudes towards the Australian mining town women she studied. She states that the liberal discourse of the 'client/individual/pathologized individual' and the socialist discourse of the 'proletarian/militant/supporter cum leader' both deny the multiplicity of subjectivities actually existing among the women in mining towns. The aim becomes one of helping to create alternative discourses that would provide more interactive, dialogic and politically enabling ways of interacting with – and understanding the identities of – the mining town women.

In a conclusion I attempt once again to tie together some of the principal themes that run throughout the book. I also point to a few areas of potential danger in the focus on difference and situatedness rather on identity and universality. Nevertheless, I enthusiastically endorse the project of empirical research and reflection based on the idea of always already contested relations of gender and sexuality which are embodied, engendered and embedded in place and space.

NOTES

1 Along with Young's reference here to women I would add others marked as 'Other'.
2 On third space and 'in-between spaces' also see Bhabha (1994).

PART I

(RE)READINGS

FEMINIST THEORY AND SOCIAL SCIENCE

New knowledges, new epistemologies

Linda Martín Alcoff

New feminist work in geography has (at least) two disciplinary discourses within which it participates: geography and feminist theory. This chapter offers an overview of feminist theory concentrating on those aspects that have been particularly relevant to work in the social sciences. And feminist theory in turn needs to be understood in its relationship to a larger historical context of academic enquiry, in order to reveal something about both its past and future. Needless to say, in taking an incoherent amalgam of diverse work and artificially producing a coherent historical narrative called 'feminist theory' I have necessarily left more things out than I can cover. I have decided to focus more on ideas than on specific people, on critiques of reason and methodology more than on substantive explanatory theories, and on Anglo-American theory more than any other. So what follows is decidedly only a part of the story.

THE CRITIQUE OF REASON

By the end of the eighteenth century, philosophy had discovered, with the help of Kant, that reason, knowledge and in fact philosophy itself was limited by the intellectual and perceptual attributes of man, that our reasoning capacity provides as much a reflection on us as a window onto the world. Indeed, as John Donne might have put it, human knowledge works more on the model of a drawing compass, whose fixed foot leans and hearkens after, but remaining always connected to the foot which strives to reach beyond. Man organizes and shapes his world, conferring on it meaning and intelligibility, and thus man is a constitutive condition of all knowledge. Philosophers continue to struggle with the implications of this idea, perhaps the most important of which is that, as Martin Heidegger said, the world which is the object of our enquiry is a world whose reference points all point to us, a lived world, and not a world in itself, or a world indifferent to human projects and concerns.

In the nineteenth century, with the help of Hegel, philosophy began to understand that knowledge and reason are also embedded within and marked by history, and thus temporally located or indexed, and unable ever to surpass completely the horizon of their historical era. Neither philosophical puzzles nor their solutions have a timeless reach, and in fact many resolutions develop only through the historical evolution of social change. Marx identified a further fundamental qualifying condition for philosophy in material power, which he defined as forms of labouring practices and relations of production. After Marx, reason and knowledge were understood to be mediated by class, situated in particular economies and permeated by an ideology that obstructed the self-criticizing project Kant had initiated. After Marx, philosophy could no longer be entrusted to discern and correct its own errors; it required external critique from other disciplines in order to reveal its ideological content.

Nietzsche and Freud also contributed, of course, to the undermining of the rigid demarcation between abstract reason and the desiring body, with Nietzsche arguing that the body is a fundamental source of all human thought and argument and Freud arguing that the rational ego maintains its autonomy over a-rational desire only temporarily.

In the (late) twentieth century, I believe it will in the future be said, philosophy began to discover that its categories of reason and knowledge are marked by sexual difference. Feminists have argued that these concepts of reason and knowledge, as well as those of man, history and power, are reflections of gendered practices passing as universal ones. What feminist theory has inserted into the critical project of our era is the sexually specific body, as a mediating element of knowledge, a constitutive component of reason, and a condition of the right to know. Let me emphasize from the outset that this is not to say that women have our own innate reason, or that truth is relative to one's gender, but that, in other respects no less important, reason is indeed 'male'.

To say that 'reason is male' is more than simply to say that men have been biased against women's capacity to be rational. It is to say that reason has been defined in opposition to the feminine, such that it requires the exclusion, transcendence and even the domination of the feminine, of women and of women's traditional concerns, which have been characterized as the site of the irreducibly irrational particular and corporeal. Moreover, as Genevieve Lloyd has pointed out, 'femininity itself has been partly constituted through such processes of exclusion' (see Lloyd 1984 esp. p.x; see also Le Doeuff 1987, esp. 'Long hair, short on ideas'). The woman who reasons, declared Kant, might as well have a beard. It is our irrational, intuitive and emotional characteristics that both define us as female and make us capable of affirming men's 'essential' superiority.

MALE MINDS/FEMALE BODIES

The major factor in this masculinist formulation of reason has been mind–body dualism. From the time of Plato, reason was thought to enable the soul to reach a 'pure, and eternal . . . immortal and unchangeable' realm where truth dwells among the 'divine . . . and the wise' as Genevieve Lloyd puts it. 'The senses, in contrast, drag the soul back to the realm of the changeable, where it "wanders about blindly, and becomes confused and dizzy, like a drunken man, from dealing with the things that are ever changing"' (Lloyd 1984: 6). To achieve knowledge, Plato concluded, 'the god-like rational soul should rule over the slave-like mortal body'. In the *Phaedo* he states it even more strongly:

> We are in fact convinced that if we are ever to have pure knowledge of anything, we must get rid of the body and contemplate things by themselves with the soul by itself. It seems, to judge from the argument, that the wisdom which we desire and upon which we profess to have set our hearts will be attainable only when we are dead, and not in our lifetime.
>
> (Plato 1961: 49)

Such a view, in various manifestations, has been present throughout the history of Western philosophy, through Aristotle, Augustine, Aquinas, Bacon, Descartes, Rousseau, Hume and even Kant.[1] And needless to say, it was men alone who could hope to transcend the realm of the body, with its everyday commitments, its pedestrian passions, and its emotions shadowing the route to the Real. Women, preoccupied with the cares of the particular, more regularly reminded of their fleshly limitations, could never ascend to the plane of the universal. As Rousseau put it, 'The male is only a male now and again, [but] the female is always a female . . . everything reminds her of her sex' (quoted in Bell 1983: 199). Therefore, he advises, 'Consult the women's opinions [only] in bodily matters, in all that concerns the senses. Consult the men in matters of morality and all that concerns the understanding' (ibid.: 197). Kant, for his part, introduces interesting spatial metaphors, locating reason in the 'public space of autonomous speech' (Lloyd 1984: 67). He defined enlightenment as precisely men's ability and willingness to use their own reason in a public space, defined in opposition to a private one. For the private space, inhabited by women and children, is dominated by particular concerns and by inclinations toward others based on feeling rather than universal principle. It is only in public, the realm to which free men have exclusive access, that a universal reason can be exercised and developed. In this light, consider the dictates of the scientific method, which require intersubjective testability of hypotheses and public confirmation. Knowledge exists in public, and not in the private, domestic environments associated with women.

The maleness of reason was thus, paradoxically, both supported and concealed by this evaluative hierarchy of mind and body. When the feeling body was split from the knowing mind, only of service to the mind as a brute recorder of perceptual images, bodily differences could not be seen to play any constitutive role in the formulation of reason. The body was conceived as either an unsophisticated machine that took in data without interpreting it, or it was considered an obstacle to knowledge in throwing up emotions, feelings, needs, desires, all of which interfered with the attainment of truth. The real epistemological action was always thought to occur in the mind, which, if it could overcome the distractions of the body and discipline it to the yoke of reason, alone had the potential to achieve knowledge.

Though reason was portrayed as universal and neutral precisely because it was bodiless, this schema worked to justify the exclusion of women from the domains of the academy, of science, and from generally being accorded epistemic authority and even credibility, because women were well known to be much more subject to bodily distractions, hormonal cycles, emotional disturbances and the like. Thus Schopenhauer, in all seriousness proposed that 'in a court of law a woman's evidence . . . should carry less weight than a man's so that, for example, two male witnesses would carry the same weight as three or even four female' (quoted in Bell 1983: 279). Even Simone de Beauvoir, writing the inaugurating treatise of feminist theory of this century, agreed with the claim that women were more prone to corporeal intrusions than men, and her (in)famous solution was for women to refuse marriage and motherhood.

> The female, to a greater extent than the male, is the prey of the species . . . in maternity woman remained closely bound to her body, like an animal. It is because humanity calls itself in question in the matter of living – that is to say, values the reasons for living above mere life – that, confronting woman, man assumes mastery. Man's design is not to repeat himself in time: it is to take control of the instant and mould the future. It is male activity that . . . has prevailed over the confused forces of life; it has subdued Nature and Woman.
>
> (quoted in Lloyd 1984: 100–1)

As long as the body and the realm of the domestic were seen as obstacles to reason, cognitive achievement and, indeed, freedom, women who sought equality had to establish an ability to transcend the body and its distractions. So even in 1970, Shulamith Firestone was advocating test-tube reproduction in order to free women's bodies from the material ties that oppress us. And some career feminists still today pursue a strategy of showing that women in management can be just as cold, detached and unfeeling as men. It is precisely for this reason that Genevieve Lloyd argued in 1984 that a feminist project determined to gain for women the realm of the 'mind' will never work to over-turn male supremacy. We cannot simply remove women from the sphere of the 'body' and claim for ourselves the sphere of the 'mind' and 'reason' when these latter concepts have been constructed on the basis of our exclusion. Such a strategy would only participate in the violent erasure of women, continuing the valorization of the masculine as the only gender that can achieve full humanity. Thus Lloyd (1984: 107) warned that,

> the confident affirmation that Reason 'knows no sex' may likewise be taking for reality something which, if valid at all, is so only as an ideal . . . If there is a Reason genuinely common to all, it is something to be achieved in the future, not celebrated in the present.

The academy today continues to be dominated by this conceptualization of knowledge and reason. Knowledge requires public confirmation, universality and a demonstrable transcendence of emotion and commitment. Knowledge must be capable of being expressed as an immaterial abstraction, beyond the irreducible concreteness of the particular, and can only be achieved in the public domain, among men, primarily through the aggressive interplay of adversarial discourse. Knowledge does not occur in private, it does not occur within the context of loving relationships, and it cannot occur where research is guided by political commitments.

By the early 1980s, feminist theorists thus began to recognize that they needed to develop a better account of the relationship between reason, theory and bodily, subjective experience. To paraphrase Rosi Braidotti, we need to

elaborate a truth which is not removed from the body, reclaiming [our] body for [ourselves] . . . [We need] to develop and transmit a critique which respects and bears the trace of the intensive, libidinal force that sustains it.

(Braidotti; 1991: 8)[2]

If women are to have epistemic credibility and authority, we need to re-configure the role of bodily experience in the development of knowledge.

In light of this, a new conception of reason has begun to be developed within feminist theory. This represents an ambitious undertaking, which is still in its early stages, but it starts from the following premises: (a) the mind is not in fact separable from the body; (b) from which it follows that the mind has never been separated from the body; (c) from which it also follows that our dominant conceptions and ideals of reason have been connected to bodies, have been expressions of bodily concerns or needs and reflections of embodied ways of being, and have had other interesting relations to the body that we have yet to discover; (d) and which also suggests that we need to rethink the entire opposition that has been drawn heretofore between reason and its 'Others', Others which all, in one way or another, have to do with the body: as rhetoric, irrationality, dreams and so on.

The project to 'reinsert' the body is not, of course, totally new. Marx inserted the labouring body into philosophy, Nietzsche reminded us of the body that feels and needs, and Freud insisted that the desiring body is a ubiquitous element in all human thought and practice. Feminism simply pointed out that these bodies are both sexually specific and sociocultural, that they are inscribed by power, and that the Kantian 'man' who conditions all knowledge is, indeed, a man, and not a woman.

RECONSTRUCTING REASON

I mark the development of contemporary feminist theory as beginning from this point, where sexual difference as a bodily, material, corporeal manifestation becomes a player in the critique of reason. I understand feminist theory today as pursuing the implications of this claim, and the effects of sexual difference on the methodologies and existing knowledges in every academic discipline, as well as exploring what a new vision of knowledge might look like without mind–body dualism, without a pretence to neutral universality or an erasure of sexual difference. I know that such statements may raise the red flag (or is it the red herring?) of essentialism, or be taken to imply that feminism is thus committed to the eradication of reason in favour of a celebration of its Others (maybe a Wicca fest in the forest). But I am not proposing a reasoned defence of irrationalism, nor advocating that a female reason should replace the male one. The feminist critique I have been describing holds out for the possibility of that future reconstruction referred to by Lloyd that would repair the split between reason and its material basis, though whether this can be accomplished with a universalizable reason left intact is still up for debate. Feminist theory is pursuing this possibility by advancing two complementary projects: first, a reactive project to critique existing theories and notions of theory itself, as well as to identify the ways in which sexual difference has both constituted and been constituted by existing knowledges, and, second, a constructive project to develop alternative theories and theoretical norms, not simply as a reversal but as a strategic transformation.[3] Both projects are pursued in the chapters included in this volume.

FEMINIST FORAYS

These projects of course evolved historically, and not just out of the critical philosophical traditions from Kant onward that I traced out earlier. In the 1970s women began to inhabit the US academy in numbers previously unthinkable. Before that time, daughters were rarely given as much money for their education as sons, much less the encouragement and support a graduate education requires, and it was not illegal for graduate programs to discriminate against women in handing out fellowships and assistantships. Thus the first female philosophy professor I ever had, in the second year of my graduate work, was there only after having overcome the trauma of having a fellowship taken away from her because, as her graduate director told her, spending money on a woman was a waste. Before I could finish my own degree, I had had a professor try to undress me in a hallway, had another professor tell me I should be at home with my children instead of at school, and was refused an incomplete I requested for missing two weeks of class due to the children's chickenpox, on the grounds that my children should not interfere with my graduate work. In those days women who were able to enter the academy by hook or by crook knew they were an unwanted, alien species, there subject to good behaviour, and only to the extent they could establish a superordinate proficiency at the traditional disciplinary procedures and standards. To challenge the presuppositions of those standards was a luxury available only to the more secure.

But in the 1970s, things began to change. Not only were women entering academic professions at a brisk pace, but there was also a women's movement both on the campus and outside of it, agitating for women's rights and critiquing male supremacy. Armed with the power of increasing numbers as well as social ferment, women in the academy began doing different kinds of academic research, research that included women, focused on women's lives and validated women's experience. What followed was an explosion of research, the development of women's studies programmes across the country and a fresh take on stale topics which infused new life in many areas of study.

Feminist forays into the social sciences thus *began* with the desire to develop more adequate theories about women, to uncover the hidden history of women, an anthropology of women, a less biased psychology of women. But it quickly became apparent that women's experiences could not in fact be included in pre-existing theoretical frameworks. For example, in political science, 'politics' was defined as what occurs in government, as influenced by business and labour unions. Given this definition, women could not be included as significant political agents. Feminist political scientists then had to redefine politics itself to include anything concerning relations of power and privilege; on this definition, 'personal' relations in the home came into view, as did schools, religious institutions and other places in which women played a significant role (see Keohane 1981).

A similar story occurred in economics. It was not until the 1960s that economists began to perceive the long-standing wage differentials between men and women as a problem worthy of analysis. But the traditional variables used in neo-classical economic theory could not account for such differentials, since they were not based on a difference in education, length of employment, productivity or the profitability of women's work *vis-à-vis* men's. Thus market forces as conventionally defined were not the key factor producing women's lower wages, which suggested further that, if market forces had not determined women's lower wages, then market forces could not be relied on to increase them. The study of women's wage differentials thus called into question laissez-faire economic theory (see Barrett 1981).

There were many research areas concerning women's lives in history, sociology, psychology and anthropology that had never been explored, most often because role-based male and female activities were taken to be gendered by nature and therefore in no need of explanation. And female-specific activities such as gathering were consistently devalued and treated as unimportant in the evolution of culture or the development of society. Women's activities were assumed to be guided by natural instinct, subject to few alternatives, ultimately uninteresting, and thus unnecessary to analyse. So women were not studied, and as a result many of the key concepts used in these disciplines were entirely based on data collected from male experience: for example the concept of the adult healthy body was based on the stability of adult male bodies, which automatically rendered pregnancy a 'disability' since such radical bodily transformations never beset the adult male body unless it became diseased. And the normative concept of a person used in psychological profiles was based on male psychological characteristics, which produced a standard that identified women as 'overly' dependent and emotional, and therefore very often 'crazy' and dysfunctional. And such key concepts as class, race, community, socialization, social control and social conflict were conceived in terms of male relationships and activities, with the result that in anthropology, for example, accounts of class and state formation neglected to mention the role, opinions or actions of women, and in sociology, social conflict was analysed as if women played no role at all.

Thus feminist researchers in these fields quickly surmised that the inclusion of women's lives and activities was going to exact a price, that it was going to affect the work of their discipline in its entirety. We could not simply 'add women and stir': the studies of women change fundamentally the way in which every discipline must delimit its subject matter, its methodology and its very self-understanding.

GENDER IDEOLOGY

Feminists also began to argue that what needed to be done was not simply to study women, but to study men in a new way, such that gender roles and sexual relations would be problematized, in order to explore how gender systems are constructed and to question the naturalness of gender itself as a binary opposition. I have avoided defining these terms – 'gender', 'sex' or 'sexual difference' – because their definition is precisely one of the fundamental topics still heavily debated. Despite the disagreements, however, feminists are in consensus that these terms need to be de-naturalized, that they need to be understood, in other words, as capable of being other than as they are or appear to be in today's societies. And feminists are also in agreement that gender ideology permeates deeply into every other ideological system, discourse, institution and set of practices. One need not study women or gender roles to find gender ideology. I have already begun to suggest how that is the case in respect to reason; I will try to show more ways in which it is the case in a moment. But first I want to briefly explain some of the ways in which feminists today are accounting for gender ideology.

Any careful reading of Marx will show that, for him, ideology is not simply a set of ideas: it is a form of practice, or a system of practices, which are connected to systems of social meanings and commonly held beliefs. It is not a set of practices determined by nature, but one that arises within particular social and historical events. *And* it is a practice that involves domination and exploitation, even while it obscures the mechanism and sources of that domination.

Gender ideology, on this model, is that set of practices which organizes, regulates and defines relations between men and women, including sexual activity, reproductive activity and gender-based roles of all types. But, more surprisingly, gender ideology also works *to produce gender*, or masculinity and femininity. Anthropologist Gayle Rubin, building on Lévi-Strauss, developed a theory that sexuality is not determined by anatomical genitalia but works to transform biological sexuality into 'sex as we know it – gender identity, sexual desire and fantasies, concepts of childhood' as well as economic and social roles (quoted in Flax 1990: 144; see also Rubin 1975). Similarities between the sexes are repressed; sexual desire is channelled exclusively toward members of the opposite gender; and female sexuality is constrained so that men can exchange and distribute women among men. Rubin also borrows from psychoanalysis in order to describe 'the residue left within individuals by their confrontation with the rules and regulations of sexuality of the societies to which they are born' (quoted in Flax 1990: 145).

In this sort of analysis, then, gender is placed squarely in the category of culture. It is not nature that must be transcended to achieve liberation; rather, it is a cultural system that we ourselves have set into motion. Gender systems are not the legacy of nature; they are the legacy of a power struggle.

Most feminist theorists today adopt something similar to Rubin's account, in which gender identity and sexuality are taken as social constructs rather than natural attributes, however remediable. Why, then, you might wonder, did I begin this chapter claiming that sexual difference should be taken as having as fundamental a status in marking reason and knowledge as history and power? Given Rubin's analysis, it would seem that, unlike history and power, sexual difference is something we must strive to overcome rather than reify as a standing feature of human life.

This points us to one of the most important current debates among feminists, and another of the ways in which current feminist theory differs from its initial, early 1970s self-understanding. Rubin's theory, though widely influential, was criticized on the grounds that the sex/gender system she invokes retains too naturalistic an account of biological sex itself. As Jane Flax (1990: 146) puts it, Rubin's opposition between the realm of the biological/natural (sexuality) and the social/cultural (gender) 'may itself be rooted in and reflect gender arrangements'. Conceptualizing the biological realm as natural conceals the way in which, even at the infantile stage, biological development is relational and therefore social, and subject to socially produced alteration. The point is not to replace the word 'nature' everywhere it appears with the word 'culture', which would reverse the opposition without subverting it. The new phenomenologies of embodiment written by feminists such as Iris Marion Young and Sandra Bartky reveal that embodiment, meaning here simply human lived experience, is simultaneously natural and cultural. An overemphasis on culture implies infinite plasticity; human experience belies that claim. To see every feature of human experience and practice as simultaneously natural and cultural is to see that every arena of human life is imbued with social meanings and subject to cultural analysis and criticism.

Sexual difference, then, must be inserted within history as well, rather than assumed as a universal which transcends historical location. There are differing accounts of sexual difference: differing accounts of its meaning, its origin, its degree of plasticity, its implications for theory. But the point I am trying to make is that few if any feminist theorists today continue to uphold the earlier feminist belief that we can now begin to theorize a 'genderless' world, or to produce theories as if their founding assumptions and method-ologies were gender-neutral. Feminist theorists have found that the very terms by which we articulate our own political goals – freedom, selfhood, empowerment, truth – have been constituted through concepts of sexual difference just as we saw that reason has been. The fact is that sex and sexism are much more deeply rooted in our society, our language and our ways of thinking than we initially understood.

I want to develop this point further by way of returning to my narrative about the work of women in the social sciences. Women theorists who were trying to study women or subjects having to do with gender initially faced an institutional dilemma. In order to gain approval within the reigning dictates of the academy, they had to demonstrate a capacity to be 'objective', that is, neutral researchers, detached from their objects of study. This required them, in effect, to disavow being women. On the other hand, they could maintain an identification with their subjects, women to women so to speak, but then lose their credibility and along with it, their grant funding, promotions and even their jobs. This recurring dilemma, which was repeated throughout different departments, led women theorists to see that demanding the inclusion of women was insufficient. And this development within the academy correlated to a corresponding development outside it: where the demand for equality began to give way to the demand for autonomy and self-determination. If the demand for equality implied an acceptance of given standards and sought an equivalence with a given norm, the demand for autonomy retained the right to reject that norm.

So what I have called the move toward a recognition of sexual difference can also be called the emerging struggle for both a political and a theoretical autonomy for feminist theory and enquiry, a demand which manifested itself in the academy as a deeper methodological critique of the roots of sexism and patriarchal assumptions in all existing domains of knowledge. A practical motivation for this was the fact that rational arguments were not working, either in the streets or in the universities, to significantly empower women. It was clear (to us at least) that male supremacy was an irrational practice, based on desire, emotion and wishful thinking more than fact or logic, but (some) men would get quite histrionic if their dominance was challenged even with the most patient, careful and well-documented of arguments. And they would even claim that reason was on their side.

Feminists then, as I have said, realized that the battle would not be won by remaining on the plane of reason, at least not as it was traditionally understood. We were then motivated to look for alternative accounts of how reason is structured: from discourse theory, structuralism and post-structuralism; as well as accounts of the so-called 'Others' of reason: from psychoanalysis to rhetoric and literary forms of analysis. Feminist theorists were motivated to uncover the workings of sexism and patriarchal assumptions wherever they were at work, and to problematize gender as a contingent rather than necessary system of practices that permeates every other system of practices in our society. The point was not to uncover the root of male evil, but precisely to discover why even well-intentioned men, as well as women, had difficulty combating and even at times perceiving the effects of misogynism. I want to outline just two more of these deeper ways in which sexism operates before I conclude with a brief look at some of the new problems feminism faces currently.

FEMINIST EPISTEMOLOGY

First, feminist work in epistemology. The very terms 'feminist social science' or 'feminist science' were perceived, of course, as oxymorons: one cannot have feminism as a political commitment connected in intrinsic ways to scientific research without violating one of the central dictates of the empirical method.

Political commitments and values could only be either irrelevant or work as obstacles to the achievement of objectivity and truth. Out of this conflict feminists began to work in epistemology.

Most epistemologists and philosophers of science of all stripes have today given up on the notion that knowledge occurs primarily through an individual's passive observation of reality, the sort of picture Adorno characterized as 'peephole metaphysics'. What's wrong with this epistemological account is precisely its metaphysical picture of the knowing process: an isolated individual, encumbered only by a language which is thought to be a neutral medium, observing nature as if through a keyhole or from an observation post. This view has been characterized as a sort of observatory model of science, predicated upon taking astronomy as the paradigm case. The astronomer must of necessity rely on observations at a distance of phenomena over which she or he has no control and no ability to manipulate. But this is, of course, a very restricted model of enquiry, and in fact ungeneralizable. The more likely general model would be something like a laboratory, in which collectives of hierarchically related individuals engage in projects in humanly created spaces, projects which are themselves determined by a variety of forces beyond any single individual's choice. On this model, knowledge is the product of cooperative human interaction with an environment. And it quickly becomes obvious that the nature of that interaction – its inclusiveness, the degree and nature of its democracy and reciprocity, the quality of its cooperation – will have a substantive impact on the knowledge produced.

To see why this has such a substantive impact, add to the above claim the fact that there are metaphysical and normative background assumptions operative in (and even indispensable to) all forms of enquiry. Feminist work has expanded on this to say that the collection of assumptions and values any given individual works with is not happenstance but can be connected in interesting ways to that person's social, cultural and political identity or location, and that most of the operative assumptions and values in enquiry are therefore group-related. This is the meaning of the often misunderstood feminist claim that there exists a relationship of *partial* determination between theories and the social identity of theorists (i.e. their gender, race and class, for instance). It is important to note that in its general form this claim places no necessary primacy on gender over culture, race or other such categories as the principle component of identity.

To accede to this claim (that there exists a relationship of *partial* determination between theories and the social identity of theorists) does not require us to hold that scientists or philosophers as a group have intentionally promoted their own privilege or have been uniformly unwilling to use the available unbiased methods of argument. For example, in Helen Longino's account of theory-choice in her recent book *Science as Social Knowledge* (1990), an account which applies in its general terms to any form of enquiry, she argues that background assumptions which contain metaphysical commitments as well as contextual values enter necessarily into the process of justifying claims to know. The influence of these assumptions and values cannot be restricted to the so-called 'context of discovery' because they have an important impact on the formulation of hypotheses, which hypotheses are taken to be plausible, the kinds of analogies and models which get seriously entertained, and the determination of the kind of evidence considered necessary or sufficient to justify theories.

When cultural and social particularities affect enquiry in such intrinsic ways, and when group-related background assumptions and normative commitments that are operative in science become restricted to a small subset of the population, the resulting effects on scientific theories should come as no surprise. So we have theories about the genetic determination of rape, molecular biology models based on corporate management structures, and theories that trace women's resistance to male dominance to PMS. There should be no surprise here. The problem is not that researchers are not following the dictates of the scientific method. The problem is that the reigning scientific method cannot reveal – much less incorporate – critical reflection on group-related assumptions. When the entire research group shares a set of assumptions, it becomes invisible. The demand for political neutrality, then, works to repudiate precisely those commitments to the democratic inclusiveness of enquiry which could improve science's ability for self-correction (see Harding 1991).

FEMINIST READING STRATEGIES

Neither science nor reason works entirely through logical entailments between factually based claims. Part of the way in which models and hypotheses are judged as worthy of experimental pursuit involves coherence, analogy and metaphor. And surely the most ubiquitous metaphor of all involves gender. Even in English, which does not gender its nouns in the way that French and Spanish do, there are many gendered associations: mother earth, boats, ships and hurricanes are female, as is the sea, justice and so on. And there are dozens of cliché phrases in the academy such as 'the penetrating argument', 'the thrust of an argument', a 'rigorous critique', 'erect a defence', a 'seminal work'. If one is in doubt that phallogocentrism exists, one need only read Saul Kripke, the influential philosopher of language, for whom the ultimate, fixed, and essential meanings of words are always determined by what he calls 'rigid designators'.

Metaphors help to elucidate meaning through making an association between two different things; they thus perform an act of mediation, which changes the term thus mediated. Hegel argued that, in the master/slave dialectic, the slave's subjectivity is mediated through labour, that in creating a new object a new sense of self is produced in the slave. Moreover, the slave's recognition of the master's status as master mediates the master's own sense of self, and thus the master's relation to himself. De Beauvoir carried this idea forward to male/female relations: a man is made to feel stronger, larger, more intelligent, when paired with a relatively smaller, weaker, dependent woman. In this way mediation transforms the self, and gender dimorphism is selected for in the species.

Metaphors in language can also perform such an act of mediation, by structuring an unstructured conceptual domain. We say of an idea or a fashion: 'It's hot!' thus transferring the semantic relations between hot and cold to those between ideas or clothing styles. Hegel himself structures the relation between men and women through the analogy of animals and plants. Guess who gets to be the tomato.

The concept of woman mediates the relations between man and his Others – other men, nature, his own self. This is not a reciprocal relation: women are defined in reference to men, as helpmates, wives, mothers, caregivers of men. Men are not defined to the same extent by their relations to women. Thus men do not figure as metaphors so often. Eva Feder Kittay (1988; see also 1987) has developed a typology of such gendered metaphors. First, man locates himself in his domain in relationship to woman in her domain, but always according a greater value to the male activity. Thus Socrates the philosopher portrays himself as a midwife, but bringing forth universal truths rather than particular babies. Second, man locates himself in his object world through a relation to women. Thus the city of Babylon is said to be the great whore, there for man's delight and temptation, and nature is, of course, a woman, trying to hide her secrets under her skirts. And third, and most obvious, woman mediates the relationships between men, establishing their status *vis-à-vis* other men; the beautiful model on the arm of a high-powered man is there for other men to see. Gang rape too is an act establishing bonds between men via the domination and subjugation of a woman.

Kittay concludes from this that woman's usefulness as metaphor depends on the difference and lesser status of our activities. Our empowerment and our equal participation in male domains will make us less useful for the mediating function. In part I bring this up to flesh out some of the ways in which Longino (1990) argues that models and metaphors which carry political implications work within enquiry to make arguments persuasive, hypotheses plausible, and to provide a coded discourse which can make us comfortable (or not) with other enquirers. Kittay's analysis also can help us understand why what might appear as trivial linguistic conventions (the subject of what today is called the pc wars) can have significant political meaning and effects.

But this analysis of metaphor should also reveal that depictions of women, of our nature, our role and our feminine essence, have the power of holding in place an entire social system of relations and practices: hierarchical relations between men, practices involving our environment and the very concept of man itself. Here perhaps, is the key to understanding the resistance feminism has encountered.

IMPERIALISM WITHIN

If Western knowledge and reason have been marked by the social location of its theorists and its institutional location within the academy, feminist theory has likewise been marked and limited by its location largely within the West. 'Woman' is itself not a coherent concept. Just as woman works to mediate man, so too do race and culture mediate gender to produce various formulations of femininity which do not add up to a coherent whole. Gender or sex cannot be taken as more central to concepts of the self than race; this may be true at some historical moments and not at others, but gender cannot be taken as conceptually primary. Which group of women's experiences should be taken as the paradigm case from which we generalize to the whole? Feminist theory has committed analogous errors to androcentric theories in insidiously privileging the position and experiences of academic European American women.

Western feminist knowledges of non-Western women often define those women not as constituted by their social context, but as prior to that context. Non-Western women are lumped into a monolithic category, defined in terms of their victimization, and rarely seen as self-interpreting beings that resist male supremacy. Today Western feminist theory is struggling with the development of new theoretical constructs which will not reinstitute such imperialist concepts as the 'historical progression of cultures' nor use non-Western women to mediate the self- identity of Western women and produce pleasing reflections of them as liberated, enlightened and in control of their lives.

And within Western feminism itself the debate continues about how to negotiate between constructing woman as victim, in light of the reality of male domination and violence against women, and constructing woman as freely choosing agent. How can the victimizations of women be brought into the light without reifying women's victim status or denying women any agency over their own lives? Katie Roiphe and others who attack feminist theory from the outside (for allegedly seeing all women as victims) may be surprised to learn that this debate has been going on within feminism for a good ten years, sparked initially by feminist historians' work on Nazi women and slave-owning women: women who, however socially constrained their lives, had to share responsibility for the brutal victimization of others. Feminists today are struggling to develop new concepts of the self which can make sense of these complexities, and to produce social ontologies which are not based on the manichean binarisms of victim/victimizer, oppressed/oppressor.

However, unlike Roiphe, the critics of Western feminist theory which have identified its colonialist baggage, theorists such as Aihwa Ong, Chandra Talpade Mohanty, bell hooks and Gayatri Chakravorty Spivak, are not calling for Western feminism to disappear but for it to develop a greater self-consciousness about the dangers of its own institutional and cultural location. They have called for feminist theory to theorize heterogeneity in place of binaries, so that the complicated relationships between men and women of oppressed races and nationalities might be more accurately described; to see women as self-interpreting subjects, and 'resist the tendency to write our subjectively defined world onto an other that lies outside of it'; to change the patronizing project of 'what can I do for them' into a more egalitarian 'speak to, learn from' relationship which recognizes that others' world-views are not merely to be corrected (Ong 1988). Imperialism cannot be overcome by producing more descriptive works on 'Other women', or by cataloguing the exotic. Rather, feminism must explore the systems by which it itself produces 'Otherness'.

THE ETHICS OF SEXUAL DIFFERENCE

Within European feminism, Luce Irigaray's work is perhaps most emblematic of this new direction for feminist theory. Her project is to create a theory and an ethics which can be grounded for the first time, not on sameness or universality, the ability to apply a method or a principle to all, or to maintain a pretence of neutrality, but a theory and an ethics that is based on difference at the most fundamental level (see, for example, Irigaray 1993). Such a requirement follows necessarily from the repudiation of mind–body dualism, since the recognition of a rich corporeality, unable to be abstracted from its particular specificity, demands the recognition of irreducible difference.

To recognize sexual difference as a constitutive component of reason and knowledge, then, involves an acknowledgement of the relation between the knowing subject and the object known, and of theory's own materiality, power and desire. A universalism of theory, no less than a universal notion of woman, must be replaced by the notion of theory as a situated enquiry with a particular aim and a specific temporal and social reach. But feminist theory today seeks not to create a new theoretical norm, but to create new possibilities for discursive space, where women can be accorded the right to know. What I hope to have done with this chapter is to open a discursive space for the feminist work which will follow in this volume, a space within which academic work of a new type, of a different type, can be heard and thought within the academy itself.

NOTES

1 Linda Bell has compiled their views on women, as well as most others in the philosophy canon, in her excellent anthology, *Visions of Women* (1983).
2 See also Grosz (1993: 187–216). For a feminist rejoinder, see Lovibond (1994: 72–86). For a mediating view between these, see Alcoff (1996).
3 This description is taken from Elizabeth Gross, 'What is feminist theory?' (1987: 190–204).

SPATIALIZING FEMINISM

Geographic perspectives

Linda McDowell

ON THE DEFINITION OF SPACE AND PLACE

In these postmodern, post-positivist, self-reflexive times, when ideas about positionality, location, borders and margins are the hot words on the lips of every social and feminist theorist (Where is she coming from, man? as we used to say in the less gender-correct 1960s), it may seem curious to be writing about the need to spatialize feminism or feminist theory (there is also a larger doubt that perhaps comes with the terrain of geography – is this what we really do: 'spatialize' others' theory – add that particular focus, that added extra, that turns something into geography? But let's leave this disciplinary anxiety aside, at least for now). In contemporary theoretical debates this use of locational terminology is largely metaphorical, referring to the displace-ment of androcentric, ethnocentric 'grand narratives' from centre stage to the margins as the voices of those multiple Others, subjugated peoples of the 'Third' World, women, people of colour, those peoples labelled as mad, bad and perverse reveal the particularity of the 'universal' claims of Western theorists. It is now widely argued that the location – the standpoint – of the theorist makes a difference to what is being claimed.

But this metaphorical displacement and dislocation is paralleled by and connected to a reshaping of the 'real' world as flows of capital and labour dis-rupt associations between nations, states and borders. In a relentless search for profit, transnational corporations roam the globe overturning traditional ways of living and proletarianizing ever larger numbers of peoples, women in particular, who are often forced to move geographically, from the countryside to the town, from town to metropolis or capital city, from the Third to the First World, from the margins to the centre. Although these patterns are not new – untold numbers of people have moved over the centuries in response to famine, 'natural' disasters, slavery, economic hardship or war, but at the end of the twentieth century the scale and magnitude of dislocation and move-ment is such that it is argued that we are entering a new era – a period of space–time compression in the words of the geographer David Harvey (1989), of space–time distanciation according to sociologist Anthony Giddens (1990), or the replacement of a space of places with that of a space of flows according to urban theorist Manuel Castells (1989). The almost instantaneous transfer of information across the globe and the vastly speeded up possibilities for physical movement of people and goods has reduced the 'friction' of space, the transactional costs of overcoming distance. While Marx saw the prospect of the annihilation of space by time in the nineteenth century we seem now to be fast approaching a depthless world of surface in which all experience might soon be simultaneous: a postmodern world of hyperreality, characterized by the speeding up of time and blurring of boundaries. Thus as Foucault argued a decade ago 'We are at a moment, I believe, when our experience of the world is less that of a long life developing through time than that of a network that connects points and intersects with its own skein' (Foucault 1986: 22): a moment in which, according to some postmodern theorists, time is being replaced by space, or rather temporality by spatiality (see Soja 1989).

Before we pursue the implications for feminist theorizing of these theoretical and material displacements, of postmodern arguments about the transformation of space and time, and indeed reveal the relations of power, place and location that position contemporary 'key' theorists, I want to turn aside for a moment and address the more specifically geographical debates (in a disciplinary sense) about the theorization of space. If we hope to 'spatialize' feminist theory, or any theory, it seems important to try and define the term 'space' itself. Here, of course there is a long and familiar literature. Suffice it to say, somewhat parodying the arguments, geographers have, in the main abandoned the search for specifically spatial regularities and laws to explain supposedly spatial processes and accepted that spatial patterns are the outcome of social processes. Now this would be a trivial statement – that all social relationships occur somewhere and result in connections between people and places (although it took many years before this relational definition of space was established) – and worryingly undermining for our discipline were it not now accepted that spatial differentiation, patterns of uneven development themselves have effectivity, that is they have a constitutive effect on social processes. A recent definitional statement by Doreen Massey, a contributor to this volume, whose work has been amongst the most influential in establishing the geographers' claim that 'Geography matters!' (see Massey and Allen 1984), is particularly helpful here. While I realize she would not claim to speak for all geographers, her paper is a useful summary of the view, widely held among geographers at present, that space is relational and constitutive of social processes. Thus, she argues,

> Interrelations between objects occur *in* space and time; it is these relationships themselves which *create/define* space and time.
>
> (Massey 1992: 79; original emphasis)

Notice her insistence that space and time are interconnected – surely an unexceptional statement.

Although Massey's paper from which this definition was taken was a critique of, *inter alia*, Laclau (1990), Jameson (1991), Soja (1989), Foucault (1986) and their arguments about the definition and relative significance of space and time, it is worth recognizing that Massey too draws attention to simultaneity. Thus, she argues,

We need to conceptualise space as constructed out of sets of inter-relations, as the simultaneous coexistence of social interrelations and interactions at all spatial scales, from the most local level to the most global.

(Massey 1992: 80)

that

'Space' is created out of the vast intricacies, the incredible complexities of the interlocking and the non-interlocking, and the networks of relations at every scale from the global to the local. What makes a particular view of these social relations specifically spatial is their simultaneity. It is a simultaneity, also, which has extension and configuration.

(Massey 1992: 80–1)

And echoing Foucault, she suggests that 'Space [is] . . . a moment in the intersection of configured social relations' (1992: 81).

Surely what all these theorists are arguing, despite their apparent differences is that we have seen both a speeding up of the interconnections and new sets of configurations that connect spatial scales, particularly the global with the local in new and unexpected ways. So, in a statement paralleling Massey's emphasis on remarkably complex interconnections between the global and the local, Fredric Jameson has argued that the postmodern condition is one distinguished by, defined by, a particular set of spatial relations. Thus, he suggests,

I take such spatial peculiarities as symptoms and expressions of a new and historically original dilemma, one that involves our insertion as individual subjects into a multinational set of radical discontinuous realities, whose frames range from the still surviving spaces of bourgeois private life all the way to the unimaginable decentring of global capital itself.

(Jameson 1991: 413)

What distinguishes the world at the end of the twentieth century is the transnational attenuation of 'local' space, and this breaking of space into 'discontinuous realities' which alters our sense of ourselves as individuals, members of various groups and communities, as citizens of a nation state. Thus for theorists interested in questions about individual and social identity, whether working in the humanities or the social sciences, geographic questions, questions of location and dislocation, of position, of spatiality, and connections, are central. Further, it seems that uncertainty or anxiety is a central theme in a great deal of current work.

As Homi Bhabha (1994: 216) suggests

anxiety is created by enjoining the local and the global; the dilemma of projecting an international space on the trace of a decentred, fragmented subject. Cultural globality is figured in the *in-between* spaces of double frames: its historical originality marked by a cognitive obscurity; its decentred 'subject' signified in the nervous temporality of the trans-national, or the emergent provisionality of the 'present'.

This is a complex but provocative statement and the notion of in-between spaces is one that I shall return to later in the chapter. I wonder though whether the anxiety apparently caused by displacement and space–time compression is not gender-specific. For many women, the decentring of the local, the widening of spatial horizons may have liberating effects as well as raising new anxieties. But this is only speculation so far.

Perhaps now, however, we are ready to start, to begin an investigation of the effects of the enjoining of the local and global on gender relations, on women's lives in particular, on the ways in which we theorize, as feminists, prospects for changing the world for the better. But two final prefatory, or qualifying, statements are needed.

First, while the definitions of space I have drawn on help us specify our focus and establish the crucial significance of interconnectedness, in my view they remain at too high a level of generality. Spatial configurations, connections between places, are significant only in the context of a specific question or investigation of particular sets of relationships. So, for example, we might investigate how and why patterns of world trade and debt position countries/ localities and individuals in particular sets of power and dominance.

Thus as Massey (1992: 81) insists 'Space is . . . a complex web of relations of domination and subordination, of solidarity and cooperation': what, in a wonderful phrase, with its implicit reference to Haraway's (1991) notion of a geometry of difference, she has elsewhere termed 'power-geometry' (Massey 1993).

Clearly, depending on their position in the social structure, people are differentially located in space, with differential abilities and opportunities to overcome what geographers refer to as the frictional effects of distance. While we are all affected by the radical transformation of local and global relations outlined above, by the power of multinational capital and global telecommunications, there are radical inequalities in the spatial spread of individuals' lives. For some, the network of points or skein referred to by Foucault above is a tightly constrained local pattern, the skein, with its wonderful woolly metaphor, is a trap, whereas for others the interstices of the network are separated by enormous distance and the connections are paths to greater freedom, an internet in cyberspace perhaps, rather than the homely skein of knitting wool. And, as feminist geographers have documented in numerous case studies in the last two decades, it is often women who have the most spatially restricted lives (Brydon and Chant 1989; Hanson and Pratt 1995; Katz and Monk 1994; Little *et al.* 1988; Momsen and Kinnaird 1994; Momsen and Townsend 1987; Tivers 1985), trapped in the net rather than free in (cyber)space.

Geography, to my mind, is defined by a middle-range focus, and by its comparative nature; the focus is on connections, looking at the links between processes and people at a range of spatial scales from the local to the global, and the ways in which these scales are themselves fundamentally interconnected – in the current clumsy but evocative phrases, what we are interested in is aspects of, the differential effects of, 'glocalization' or global localism.

The second point is a separate but related one – in the focus on change and simultaneity, on the overcoming of space and distance, it is important that we do not forget permanence, solidity, meaning and symbolism, what we might refer to as attachment to place. For space is not just a set of flows, *pace* Castells, but also a set of places, from a home to national territories, with associations and meanings for individuals and groups. Here the distinction between space as relational and place as a location, as a fixture, the geographer's 'sense of place' or 'genius loci', what the literary and social critic Raymond Williams termed a 'structure of feeling' centred on a specific territory is a useful one. This is not to argue, of course, that the meaning and symbolism of place is unvarying – this is clearly absurd – but to emphasize that spaces and places are not only sets of material social relations but also cultural objects. Thus we must investigate not only patterns of flows but also the meaning of place, of place as absolute location, and of place as stasis, albeit with varying boundaries. If the last decade has taught us anything it is surely that the meaning of territory has a continued significance in the contemporary world as peoples variously fight over their attachment to place – in the former Yugoslavia, in the former Soviet Union, in inner cities all round the world.

I want now to turn from this definitional focus to specifics and to consider the ways in which the phrase 'spatializing feminist theory' might be interpreted. I have chosen to do three things in the second part of this chapter – addressing, first a question about the location of theorizing; second, examining some of the material consequences of globalization and migration for the attachment of identity to place; and, third, examining the prospects for the construction of a feminist politics that reaches across boundaries. In all three cases I shall focus on space as a metaphor as well as a set of material social relations. It is only the second of the three sets of issues that is, in a particular disciplinary sense, geographical (and even here the questions are new and unrecognized as yet in more conventional geographic discourse). But no matter. As the chapters in the rest of this volume show, feminist geographers have long dealt with issues and questions that have challenged the very conception of our discipline. Like feminist scholarship in general, the excitement and achievements of the last two decades or so have lain in the radical challenges to the nature of disciplinary knowledge per se. It is no longer possible to avoid raising questions about who is speaking, from what position, in claims to knowledge. In all three areas that I discuss below, it is hardly possible to do more than indicate some of the main lines of the argument. In many cases these will be familiar – there are now a large number of books, texts and papers by geographers and others addressing questions of location, space, place, identity and position, too numerous to list (the recent *New Formations* collection edited by Carter *et al.* 1993 is a useful starting place) – and in other cases some of the particular lines of argument that I indicate here are opened up in greater depth in other chapters in this collection.

POST-CARTESIAN POSITIONS

Contextualizing theory — acknowledging by whom and where theory is constructed and why that matters

The crisis of the rational subject has a long history. That indivisible, singular and unique human subject – the thinking individual of Enlightenment thought who saw himself as master of the universe – has been unravelling for decades. Stuart Hall, in an admirably clear essay, has traced the dislocation and fragmentation of identity and the subject through the effects of what he terms 'a series of ruptures in the discourses of modern knowledge' (1994: 120). In particular, he suggests there have been

> five great advances in social theory and the human sciences which have occurred in, or had their major impact upon, thought in the period of late-modernity [the second half of the twentieth century] and whose main effect, it has been argued, has been the final decentring of the Cartesian subject.
>
> (1994: 120)

The five great advances are:

1 *Marx*'s recognition of the ways in which individuals' actions are limited by 'circumstances not of their own choosing'.
2 *Freud*'s discovery of the unconscious in which he argued that our identities and desires are not rational but function very differently. Modern psychoanalysis has built on Freud's work to argue that self as whole and united is something which has to be learned; it is formed in relation to others and is seldom completely achieved but rather is an ongoing process involving imagination and fantasy.
3 *De Saussure*'s work on linguistics in which the subject is positioned within a pre-existing set of language rules rather than being the 'author' of statements. Words and statements always carry with them echoes of a range of meanings and so are inherently unstable.
4 *Foucault*'s analysis of individuals as the product of 'disciplinary power' which produces docile bodies. Paradoxically the surveillance and power of a wide range of collective institutions produces an increasingly individualized subject.
5 The impact of *feminism* as both a theoretical critique and political movement in the context of a wider range of social movements that are based on and draw their strength from what has become known as 'identity politics'.

While there is now a large feminist literature about the ways in which feminism has been particularly effective in the conceptual decentring of the Cartesian subject, Hall's summary of the impact of feminism is worth repeating here for its admirable brevity and completeness. There have been, he argues, five ways in which feminism has been important in disrupting the idea of a centred subject.

1 It questioned the classic distinction between 'inside' and 'outside', 'private' and 'public'. Feminism's slogan was 'the personal is the political'.
2 It therefore opened up to political contestation whole new arenas of social life – the family, sexuality, housework, the domestic division of labour, child rearing, etc.
3 It also exposed, as a political and social question, the issue of how we are formed and produced as gendered subjects. That is to say, it politicised subjectivity, identity and the process of identification (as men/women, mothers/fathers, sons/daughters).
4 What began as a movement directed at challenging the social *position* of women expanded to include the *formation* of sexual and gendered identities.
5 Feminism challenged the notion that men and women were part of the same identity – 'mankind' – replacing it with the *question of sexual difference*.

(Hall 1994: 125; original emphasis)

It is clear that in all these five areas questions about location have been central. It was feminist theory that pointed out the absence of whole areas of life from social theory, that challenged the very notion of private life that took place in particular small scale spaces – in the home, inside, in the private not the public spaces of the city. And while Hall is correct with all these five arguments, feminist theorizing has, of course, moved further and faster in the decentring than he suggests, displacing the very notion of 'woman' itself. The question of difference, which is so important, not only applies to differences *between* women, whether based on class, or on age, on ethnicity or on sexual identification, on country, region or locality, but also to the notion of the individual and the self. Ideas about a subjectivity that is decentred, fractured and partial are a challenge to the idealized rational individual. In post-rationalist, postmodern, post-structuralist or whatever we choose to label contemporary critical theories, there is a profound scepticism towards all universalizing claims about the existence and nature of powers of reason, towards ideas of progress, science, the mind/body separation and the rational subject. Instead attention is drawn towards multiplicity and differences, none of which are theoretically privileged over any other. Many feminists suggest that epistemologies must be regarded as contextual, situated and positioned, as well as temporary. As Hartsock has argued 'epistemologies grow out of differing material circumstances' (Hartsock 1990: 158): a view that elsewhere she has termed 'standpoint theory'.

While a number of feminists have regarded these conclusions with deep pessimism, believing that the deconstruction of *the* female subject undermines the basis for a specifically feminist politics, others have 'a passion for difference' (the title of Henrietta Moore's 1994 book), embracing the liberatory potentiality of a non-hierarchized multiplicity in which people are not distinguished or characterized by their difference from a white, male, bourgeois norm (their distance from this in the sense of failure) but rather multiplicity is celebrated.

The proponents of standpoint theorizing have argued the need for theories that begin from the experience and point of view of the oppressed/dominated: those whom Foucault termed the subjugated. Haraway (1991) similarly suggested that the situated view from below is likely to include a clearer perspective on the conditions of oppression than what she termed the 'view from nowhere' (Cartesian rationality). But we must include in these contextual or positioned theories, the views from somewhere, not only the relations of dominance that construct, define and oppress a particular group, be they women *per se* or older women, women of colour or whomever, but also, as Hartsock has emphasized, women's 'capacities, abilities and strengths' (1990: 158). Standpoint theories must 'use these capacities as guides for a potential transformation of power relationships'. Before turning, in the final section of this chapter, to this extremely important question of constructing a feminist politics across differences, I want to pursue the question of theory construction a little further and also raise some questions that I find troubling about multiple claims to knowledge.

Standpoint theories and feminist geometries

Standpoint theories, which are specific to time and place, have a certain appeal to geographers, and yet there is so far relatively little work by geographers that has addressed the question of how to construct contextual theories of difference in which the associations of gendered identities with place and location are seriously addressed. Instead geographers have looked to feminist theorists writing elsewhere for guidance. The work of Donna Haraway seems to have a particular resonance, in part perhaps because of her own use of spatial terminology including maps and geometries. Haraway, perhaps in a reference to Geertz uses the term 'local knowledge' or 'embodied knowledge', to emphasize the deconstruction of the mind/body dichotomy. She has argued that feminists need to develop a *geometry of difference* that allows us to consider the relations of difference in ways other than hierarchical dominance. She draws on Minh-ha's (1987) concept of 'inappropriate/d others' to refer to the positioning of people who refuse to adopt the binary identity of either 'self' or 'Other' that is offered in dominant theories of identity. Instead we need to theorize a geometry of multiple difference. Haraway emphasizes the 'hard intellectual, cultural and political work these new geometries will require' (1991: 3) but begins to sketch in the outline of a new theoretical approach. In a long passage that I have drawn on before (McDowell 1993) but which I continue to find provocative, Haraway argues as follows:

A map of tensions and resonances between the fixed ends of a charged dichotomy better represents the potent politics and epistemologies of embodied, therefore accountable, objectivity. For example, local knowledges have also to be in tension with the productive structurings that force unequal translations and exchanges – material and semiotic – within the webs of knowledge and power. Webs have the power of systematicity, even of centrally structured global systems with deep filaments and tenacious tendrils into time, space and consciousness, the dimensions of world history. Feminist accountability requires a knowledge tuned to resonance, not to dichotomy. Gender is a field of structured and structuring difference, where the tones of extreme localization, of the intimately personal and individualized body, vibrate in the same field with global high tension emissions. Feminist embodiment, then, is not about fixed location in a reified body, female or otherwise, but about nodes in fields, inflections in orientations, and responsibility for difference in material-semiotic fields of meaning.

(Haraway 1991: 195)

This concept of embodiment as a node in a set of fields variously structured by sets of social relations ranging from the global to the most intimate scale seems to me to parallel the notion of place that is common in geographical work (Massey 1992; Smith 1993), while reminding geographers that questions of identity are not solely related to the smallest scale, to the body and the home or to the community, which is where too many geographers continue to place them. If we move towards a definition of both identity and place as a network of relations, unbounded and unstable, rather than fixed, we are able to challenge essentialist notions of place and being, and of local, face-to-face relations as somehow more 'authentic' – a common strand of both modern and some versions of postmodern theorizing. Indeed, Iris Marion Young has made this specific connection in her critique of an idealized concept of local place-based community that has imbued feminist and socialist thought alike: a concept that, she argues, 'implies a denial of space–time distancing' (1990: 302).

It is to this issue of space–time distancing, communities and the implications for feminist politics that I turn in a moment, but first I want to conclude this section with a difficult question about writing and the construction of multiple knowledges.

Polyphony, cacophony or authorial dominance

It is now commonplace in what has become known as the 'new ethnography', as well as in literary theory, other aspects of the humanities and the social sciences, to argue that the author is dead. Meaning itself is fluid and contextual. The notion of a singular authoritative voice that controlled both the construction and interpretation of narratives is challenged by arguments that suggest that there are interpretative communities, that readers are differentially placed and so have different reactions to a single text.

While this seems unexceptional, a more difficult question is raised by what some term a crisis of representation, that is a crisis of writing rather than reading. Many anthropologists, and an increasing number of geographers, now suggest that the desired aim of scholarly writing is a polyphonic text, in which the multiple voices of the narrator and his (the pronoun is deliberate – these claims seem to have originated in a predominantly masculine group of scholars (see Mascia-Lees *et al.* 1989 and Moore 1994)) subjects are heard. Now, leaving aside difficult stylistic questions about how to construct a multiphonic text that is legible, it may be argued that this genre does not challenge authorial dominance. Indeed, it may even increase it and give it a new, and unacceptable, dominance. Thus while polyphony in no way detracts from the responsibility of the author to decide whom to include in the narrative, how to arrange the material, how to decide who speaks for whom and on what basis, at its best it may make this process more explicit. But there is a second consequence that is more difficult: the aim of making the author visible in the text (usually a member of a dominant group by virtue of their location as a scholar, if not always White, male and Western) has the consequence of an explicit movement of the author from the margins to the centre of the narrative. Thus authors become characters in their own stories, now explicitly rather than implicitly orchestrating the work. Now this is not an original point. Geertz (1988) pointed out, some years ago, the irony of ending up with author-saturated texts produced by those who claimed to be sharing authorship with others. More problematic, as Moore has recently convincingly argued, is the necessary continuation of older notions of an author. As she points out all these texts are based on the assumption of a singular identity – that 'the author in the text and the author of the text are one and the same' (Moore 1994: 117). And, further, this relationship is a fictive one, imaginary in that it is arbitrary and symbolic, set up in language and culturally inscribed. The author in the text is the imaginary self of the author of the text, or rather 'properly speaking, they are not two selves, but a self in process' (Moore 1994: 118).

Whatever the case, Moore has tart words to say about the importance of this crisis of representation, which as she points out is actually a *political crisis* (1994: 117; my emphasis). Moore suggests that 'it is an irony of the contemporary moment that while international capitalism and other forces threaten homogeneity, difference is on the political agenda more than ever' (1994: 117). While not wanting to read something into her words that she did not intend, I detect a note of censure here. I want to suggest, rather, that the political crisis lies precisely in the coincidence of, or intersection of, forces of homogeneity and heterogeneity. It seems to me that the current moment is such a vexed one for the construction of a progressive politics precisely because of this doubled reconstruction of spatial relationships. I attempt to expand this argument in the next section.

STRETCHING SPACE: IN BETWEEN/THE THIRD SPACE

Haraway emphasized the connections between the material and the semiotic in the webs of power/knowledge that construct our everyday lives. I, and others, have argued that geographers must link the material and the symbolic or metaphorical in analyses of the social construction and significance of space in recent theoretical endeavours, including feminist theorizing. However, a relative emphasis on either the material or the symbolic is often appropriate, depending on the questions in mind. Here, in this second section, the emphasis is on the material: on the ways in which the transition to an increasingly interconnected global economy has altered people's sense of themselves, whether they remain trapped in the same old place or literally have been transported half across the globe. For all people though, whether geographically stable or mobile, most social relations take place locally, in a place, but a place which is open to ideas and messages, to visitors and migrants, to tastes, foods, goods and experiences to a previously unprecedented extent. It is this openness that I have termed global localism here. This seems to me to better capture the different degrees of fixity than terms such as space–time compression or space–time distanciation which tend to fail to recognize the unevenness of these socio-spatial changes.

There is also another side to the recent geographic changes. We also now live in a post-colonial world in which nations are fragmenting into smaller nations and where local or regionally based social and political relations have increasing salience. Movements for the self-determination of peoples, differentiated one from another on the basis of ethnicity, language or religion, are fragmenting the world political divisions of the post-war period and producing an unfamiliar map. Thus, through a range of processes – from the uneven impact of an increasingly global capitalism, migration, war, new social movements – old boundaries are being transgressed and disrupted and replaced by new divisions. In these processes, men and women, divided or united by age, by class, or by beliefs, are differentially affected, and the links between identity and a sense of belonging to a particular territory or place are being remapped. Now this chapter is not the place to even outline the main features and implications of the breaking down of old spatial divisions and boundaries and the re-establishment of new ones. Some of these stories are told in the succeeding chapters; others elsewhere. Instead here I merely want to draw attention to the ways in which movement and migration have also forced us to rethink ideas about identity, subjectivity and selfhood by disrupting another of those significant Enlightenment binaries: in this case the division between 'the West' and 'the rest' and to draw attention to the utility of ideas of 'between-ness' – developed in the main in the work of post-colonial theorists and feminists of colour.

What globalization, the associated movement of people and capital, and the expansion of hegemonic forms of Western media into 'Others' spaces', has achieved is no less than the disruption of geographic space, or at least its definition and association with 'real space'. The movement of vast numbers of people from the formerly colonized periphery to the centre of what was once termed, without irony, the 'civilized world' has collapsed the distinction between the West and the rest, their geographic separation and the association of the cultural values of 'the West' – those Western philosophical principles (such as the Cartesian mind/body distinction) – with (in the main) the Western hemisphere. Now a multiplicity of peoples of different colour, religion and nationality make up 'the West'. These people, as Moore (1994: 132) has suggested,

> members of the British Asian and Afro-Caribbean communities
> might not readily identify with the category of the West as deployed in

anthropology, cultural studies, colonial discourse theory; with that particular set of cultural values, symbols, social structures and ways of being shored up by acts of violence and economic opportunism. And yet they so obviously are part of any sensible definition of the West; they are at the heart of the category even as they seek to resist it, transform it and educate it.

Of course what we might clumsily term 'people of colour' have always been at the heart of the definition of the West, or what it means to be Western in the sense of that being white, or Western or advanced was always defined relationally and oppositionally. It meant 'not being' whatever shifting characteristics were used to define 'the Other'. The key change now, of course, is that 'them' and 'us' now share the same geographic space – the heartlands of advanced capitalism, the metropolises of the 'First' World. The West can no longer be identified with a particular set of spaces or geographically defined people. It has become, rather, 'a discursive space, a set of positionalities, a network of economic and political power relations, a domain of material and discursive effects' (Moore 1994: 132).

This disruption of space through migration, of course, has parallels with women's position in the West, perhaps making more visible arguments from within feminism about women's awkward 'place' in the West. For women, too, were/are excluded by Western philosophical ideals, equally 'out of place' in that discursive space called the West. The long debate about the public and the private is too familiar to need rehearsal here, but it reminds us of the significance of geographical location to the construction of gendered identities. Women actively and passively, through the changing nature of their everyday lives, their position in the family, the household and in the workplace, all of which have been affected by the social relations of local globalism and its associated geographic restructurings, are challenging the gendering of space as they disrupt conventional associations between Whiteness, masculinity and the workplace, for example, between gender and political power, between femininity and accepted definitions of sexuality. At a range of spatial scales, from the most local in the home to the global scale, women and people of colour have challenged conventional assumptions about the relationships between identity, both individual and group, and location, as well as the theoretical basis of Enlightenment thought. Old associations between a place and a people, be it a community, a region or a nation, are breaking down and are being reforged at the end of the twentieth century. (It is of course, important to remember that what we tend now to refer to as 'old' or even 'traditional' relationships between place and identity were themselves reshaped by the upheavals of industrial urbanization in the modern period, as well as by imperialism, slavery and wars. And to keep in mind Anderson's (1983) work on the bonds created to construct an imagined community among people sharing the same national territory. In other words, the pace and scope of globalization may have accelerated but the phenomenon itself is not recent.)

A helpful way to conceptualize the current relationships between location and identity is in a set of interlinked concepts developed by social and cultural theorists such as Paul Gilroy, Stuart Hall and Homi Bhabha, though each takes a somewhat different approach. Gilroy, in his book *The Black Atlantic* (1993) argues that the contemporary Black English (and other migrant peoples) stand between two (at least) great cultural assemblages: between the intellectual heritage of the West and an absolute sense of ethnic difference, based on often idealized and imaginary notions of Black nationalism. Gilroy, like Hall, argues that neither of these ideals are appropriate either for understanding the identity of former colonized peoples in the West or for the basis of a political movement. Instead, it is important to theorize what he terms 'creolization, metissage, mestizaje, and hybridity' (1993: 2), although anxious about the specific terminology with, from the viewpoint of ethnic absolutism, its undertones of impurity. Hall similarly points to three contested ways of constructing a sense of self and group identity, especially a national identity, in an increasingly multicutural world. He views the first two reactions as a conservative response to anxiety and threat: the little Englander reaction of White Britons and the search for ethnic purity and a homeland among minority groups, exemplified in a movement such as Rastafarianism. The third response is more hopeful, although difficult. It lies in 'the emergence of new subjects, new genders, new ethnicities . . . [who have] acquired through struggle the means to speak for themselves' (Hall 1991: 34). These new subjects occupy new spaces – 'new regions, new communities' (1991: 34) – in which Hall emphasizes, like Gilroy, a new hybrid identity forged out of betweenness may emerge.

Bhabha (1994) too emphasizes the consequences of geographic movement for identity, suggesting that new and transitional identities, which he also terms hybrid, are emerging from mass movements and the intermixing of different peoples. In a passage full of challenge to geographers attempting to rethink their definition of space and spatiality, Bhabha suggests we are seeing the emergence of 'a third space' in the contemporary world. He suggests that

> what is manifestly new about this version of international space and its social (in)visibility, is its temporal measure – 'different moments in historical time . . . jump back and forth' [the quote is from Jameson]. The non-synchronous temporality of global and national cultures opens up a cultural space – a third space – where the negotiation of incommensurable differences creates a tension peculiar to borderline existences.
>
> (1994: 218)

What is particularly important for the argument here is the thought -provoking coincidence between Bhabha's and current feminists' arguments about gendered identities and ideas of the subject as multiple and fragmented. Here is Bhabha defining the 'residents' of the third space:

> the subjects of cultural difference do not derive their discursive authority from anterior causes – be it human nature or historical necessity – which, in a secondary move, articulate essential and expressive identities between cultural differences in the contemporary world. The problem is not of an ontological cast, where differences are effects of some more

totalising, transcendent identity to be found in the past or the future. Hybrid hyphenisations [the term is reminiscent of bell hooks's celebration of multiple hyphenated female identities (black-feminist-lesbian, etc.] emphasise the incommensurable elements as the basis of cultural identities. What is at issue is the performative nature of differential identities: the regulation and negotiation of those spaces that are continually, *contingently*, 'opening out', remaking the boundaries, exposing the limits of any claim to a singular or autonomous sign of difference – be it class, gender or race . . . difference is neither One nor the Other, but *something else besides, in-between.*

> (Bhabha 1994: 219; original emphasis)

Here we see then the coincidence between post-colonial cultural studies and feminist scholarship (Butler 1990; Leidner 1993; Moore 1994; Pringle 1989; Probyn 1994) where there is growing interest in the analysis of subjectivity and lived identity as performance, masquerade and parody. The work of constructing an identity is never complete, involving struggles and resistances as well as acceptance, pleasure and desire. Further, there is an insistence on the multiple nature of subjectivity and its construction in local or lived experiences. The sense of oneself as a certain sort of woman, defined by class, 'race', religion, age and so forth, is given meaning by the actualities of everyday experience. And this experience itself is a complex series of cross-cutting locations in which the significance of different aspects of the self varies. In these conceptions of subjectivity, location and embodiment are crucial. As Moore has suggested

the powerful symbolism of notions of place, location and positionality in contemporary feminist theory demonstrates just how much we come to know through our bodies, and how much our theorizing is dependent on that knowledge. The multiple nature of subjectivity is experienced physically, through practices which can be simultaneously physical and discursive.

> (Moore 1994: 81)

So, as geographers assert (but are relieved to find others agree) location matters!

Thus we have a coincidence between material, symbolic and discursive constructions of space, in situated theory, in imagined communities, in the social construction of different visions of space and in the performative and fictive nature of subjectivities and social relations. There is an evident drawing together of interests by a range of critical theorists in these ideas in which location and position are key concepts. For feminist, social and cultural geographers these are exciting times.

WHAT IS (A SPATIALIZED) FEMINIST THEORY FOR? CONSTRUCTING A POLITICS ACROSS SPACE. SEPARATISM?

Finally, I want to conclude with a few remarks about the prospects for a spatialized feminist politics. While postmodern/post-colonial or whatever we choose to term them theories of multiple positionalities and multiple subjectivities may have a special appeal to 'social theorists who live these complicated, conflicting and compelling differences (Moore 1994: 81) – and most feminists working in the academy surely identify with this contention – these theories also seem to make it more difficult to sustain the notion of a specifically feminist politics or indeed even to argue for the continuing centrality of gender-based analyses. Now the arguments about relativism, values and morals and the prospects for claiming 'principled positions' (the phrase is the title of a book by Squires, 1993) have been well-rehearsed by many others, including some of the most rigorous feminist theorists (Haraway, Hartsock and Fraser are names that are uppermost in my mind but there are many others). As these committed scholars have suggested, it is important that we aim for a 'social criticism that is ad hoc, contextual, plural and limited' (Hartsock 1990: 159) but one that is not disabling – as Hartsock argues, progressive social critics have to hold on to a belief in the possibility of systematic knowledge. The aim, of course, is to argue that systematic knowledge is not only possible, but that it must be contextual and local, various and diverse.

However, the aim of this knowledge construction has changed. It is not enough to assert and demonstrate difference. Those groups positioned as 'the other' by the discourses of Western science now seem to have a twofold agenda. The first aim is to dispute the categorization of their knowledge as somehow inferior, as alternative or as one perspective among many, each with equal validity. As Hartsock suggests:

> Our various efforts to constitute ourselves as subjects (through struggles for colonial independence, racial and sexual liberation struggles, and so on) were fundamental to creating the preconditions for the current questioning of universalist claims . . . out of this concrete multiplicity [we need to] build an account of the world as seen from the margins, an account which can expose the falseness of the view from the top and can transform the margins as well as the centre. The point is to develop an account of the world which treats our perspectives not as subjugated or disruptive knowledges, but as primary and as constitutive of a different world.
>
> (Hartsock 1990: 171)

The second aim is to develop political strategies to build and act on these perspectives – 'to engage in the historical, political, and theoretical process of constituting ourselves as subjects as well as objects of history' (Hartsock 1990: 170). For feminists the challenge is to begin to forge alliances between groups who are differently positioned. In many cases this will involve a hard political struggle of uniting women who are themselves divided by their class position or their ethnicity around a common series of issues, as well as working with other groups of oppressed peoples. In different circumstances, we need to ask ourselves not only what are our differences but also what are our commonalities. This political work will involve building bridges between positions, in the literal as well as the metaphorical sense. The processes of local globalization that I have referred to above mean that women in widely separated geographical spaces have interests in common. As Chandra Talpade Mohanty has reminded us, Third World women in the nation-states of 'the South' and Black, Asian and indigenous women living in 'the North' are an 'imagined community' based on Third World oppositional struggles.

> Imagined not because it is not 'real' but because it suggests potential alliances across divisive boundaries, and 'community' because in spite of internal hierarchies it nevertheless suggests a significant, deep commitment to what Anderson, in referring to the idea of the nation, calls 'horizontal comradeship'.
>
> (Mohanty 1991: 4)

And, as Mohanty, goes on to argue, reassuringly perhaps for those geographers more comfortable with a material conception of space, 'such imagined communities are historically and geographically concrete, [but] their boundaries are necessarily fluid' (1991: 5). Thus feminist activists are involved in a wide range of activities, based on alliances and the construction of communities across a range of spatial scales with differing temporalities. Within nations and communities, campaigns about legal rights, about representation, about the provision of communal goods and resources unite and divide women. Across nations, these and other issues – of war, religion, persecution, mutilation and torture unite and divide women. And the very processes of spatial globalization, of space–time compression that have had such an impact on the everyday lives of millions of people, may be used to annihilate the space and distance between them. As global telecommunications become more accessible – the fax and e-mail, for example – they may become the weapons of the weak as well as the powerful.

Spaces for women – arguments about separatism

One last point remains to be considered, which perhaps tempers the enthusiastic advocacy of the politics of alliance and coalition outlined above. As the history of 'second wave' feminist politics has taught us, it is clear that an important part of oppositional struggles is separatism: a separatism which is often associated with demands for a geographical space, or territory and the maintenance of mechanisms of exclusion and boundaries. As Hartsock recognized,

> one of our first tasks is the construction of the subjectivities of the Others, subjectivities which will be both multiple and specific. Nationalism and separatism are important features of this construction.
>
> (Hartsock 1990: 163)

This argument forces feminists to face difficult questions to which there are no easy answers. And the question of geography, or more accurately of scale, is a crucial part of the dilemmas raised by separatist movements. Thus we may support a whole variety of initiatives to create women-only spaces – from bookshops to car parks, railway carriages or art-centres, bars to peace camps – based on arguments about solidarity, comfort and safety. Indeed, I am sure that Hartsock's quotation from Bernice Ragon (who is a singer, activist and social historian), viz:

> [Sometimes] it gets too hard to stay out in that society all the time. And that's when you find a place, and you try to bar the door and check all the people who come in. You come together to see what you can do about shouldering up all your energies so that you and your kind can survive.
>
> (1990: 163)

will provoke a wry smile of recognition from most readers. It is, after all, such hard work living with 'the enemy' all the time.

But how are we to respond to the moral issues posed by what may seem like more extreme versions of geographic separatism, based on ethnic separatism perhaps, or on religious beliefs, which are often extremely oppressive societies for women? It is hard, although not impossible, work to defend certain separatist movements while denying others. This dilemma, of course, is but one aspect of the same one raised earlier in these concluding remarks: the necessity for all those who adhere to a belief in difference, in the construction of non-hierarchical local knowledges to simultaneously adhere to beliefs in systematic knowledge, principled positions and the necessity for a progressive political struggle for a less unequal world.

RE: MAPPING SUBJECTIVITY

Cartographic vision and the limits of politics

Kathleen M. Kirby

The Cartesian subject has seemed, to many contemporary theorists, the last (and first) bastion of the current political order. It has been held responsible for the atrocities of imperialism, the subjugation of women and the psychological illnesses of Western individuals.[1] Post-colonial critics and feminist critics have catalogued the ways the Enlightenment individual founded itself at the expense of others, especially Third World populations and women. Contemporary theorists are both challenging this norm's claim to exhaustively represent subjects and attempting to reconstruct this subject, where it has hardened into reality, to propose more responsive forms of epistemological and social relation.

My argument will be that the development of Enlightenment individualism was – and continues to be – inextricably tied to a specific concept of space and the technologies invented for dealing with that space. Graphically, the 'individual' might be pictured as a closed circle: its smooth contours ensure its clear division from its location, as well as assuring its internal coherence and consistency. Outside lies a vacuum in which objects appear within their own bubbles, self-contained but largely irrelevant to this self-sufficient ego. Will, thought, perception might be depicted as rays issuing outward to play over the surface of Objects, finally rejecting them in order to reaffirm its own primacy. Objects that are accepted are pulled in through the walls of the subject and assimilated, restoring the interior to homeostasis.

The Cartesian subject, the Enlightenment individual, the autonomous ego of psychoanalysis: all appear to be reducible to this same graphic schema. The 'individual' expresses a coherent, consistent, rational space paired with a consistent, stable, organized environment. Cartography, a science developing (as a science) in the Renaissance and being standardized in the Enlightenment, is both an expression of the new form of subjectivity and a technology allowing (or causing) the new subjectivity to coalesce. The form for subjectivity, space and the relation between them inspired by mapping has achieved, I would argue, a kind of popular dominance today – though all three are also beginning, according to Fredric Jameson, to 'wither away' under the pressures of postmodernism.

Briefly consulting two Renaissance texts of exploration will allow me to reveal the roots of cartography's configuration of the subject and space. Jameson's writing on postmodernism will provide a place to consider more closely the ramifications of a continued theoretical reliance on mapping. My contention is that the mapping subject, now as then, is a construct incapable of responding to many of the features of the (geopolitical) environment; that it is an exclusive structure encoded with a particular gender, class and racial positioning; that it is a structure for subjectivity unresponsive to the perspectives of many non-dominant subjectivities, particularly women. And finally, its negative qualities are an effect of the way it constitutes space. The new styles of space forming the foundation of postmodernism may offer precisely the material for building a new kind of subjectivity, one that will not leave non-dominant subjects at the theoretical and political margins.

Throughout his work, but particularly in *Civilization and Its Discontents* (1930/1961) and *The Ego and the Id* (1923/1961) Freud proposed that the structure of the self is achieved through the delimitation of an external environment, and thereby suggested that the form for the environment that the self produces will recursively dictate the shape of the self. The symbiotic shaping of environment and self that Freud observed might occur not only in the psychological developmental process that goes to form *an* individual, but in the cultural and historical process that went into shaping *the* individual, and any other form that subjectivity has historically taken. In 'The land speaks', Richard Helgerson writes, 'One hears much of the Renaissance discovery of the self and much too of the Elizabethan discovery of England . . . Not only does the emergence of the land parallel the emergence of the individual authorial self, the one enforces and depends on the other' (1986: 64). Mapping comes onto the scene to both reflect and reinforce a new way of conceiving both the subject and space.

What kind of space, what kind of subject, does mapping (per)form?[2] Cartography selectively emphasizes boundaries over sites. J.B. Harley argues that such a choice of emphasis indicates the primacy in European mapping of *ownership* (n.d. 32). One could transfer this insight into the realm of the subject by pointing out the emphasis upon 'propriety' and 'own-ness' in the 'one-ness' of the Enlightenment individual – as well as this subject's imbrication in the developing social form of capitalism. Another disposition of post-medieval conceptions of space was standardization. Harley demonstrates the tendency of early American cartographers 'to obliterate the uniqueness of the American landscape in favour of a stereotype' reflecting a European sensibility of the natural world (1988: 68). This conception of space again parallels the form of space increasingly underlying the concept of the 'individual'. Standardized 'Man', like mapping iconography, applied its own culturally specific standards as if they were indeed universal to the end that actual otherness was erased. Subjects, like places, were homogenized in favour of the generic, so that social policy based on humanism has proven insensitive to the varying needs of 'different' subjects.

Changes in the aesthetic appearance of maps testified to the growing authority of scientific discourse, which would terminate in an erasure from the map of all signs of the immediately subjective. The 'central bastions' of European mapping from the seventeenth century onward 'were measurement and standardization' (Harley 1989: 4–5). In both realms, idiosyncrasy and emotionality, physicality and specificity, are increasingly marginalized. The cultural and subjective location of mapping are elided, much as are the problematics of subjectivity for the Enlightenment *individual*. The Western subject during the Enlightenment tended to define itself by cataloguing *others* (woman, native, criminal, insane) which it opposed because it did not require definition. Each tends to project outward, to let the beam of attention play across the surrounding world, rather than turning its cognition on itself.

The space that mapping propagates is an immutable space organized by invariable boundaries, an a-temporal, objective, transparent space. Not coincidentally, the same physical qualities characterize the kind of subjectivity that we would name, variously, Cartesian monadism, Enlightenment individualism or autonomous egoism. But the relationship is not only metaphoric – one of comparability; it is also metonymic – one of contiguity. The similarity of mapped space and the mapping subject stems from the way the boundary between them is patterned as a constant barricade enforcing the difference between the two sites, preventing admixture and the diffusion of either entity. Cartography institutes a particular kind of boundary between the subject and space, but is also itself a site of interface, mediating the relationship between space and the subject and constructing each in its own particularly ossified way.

I will turn now to analysing two examples of Renaissance narrative mapping to demonstrate in more detail the consistency of cartographic space. Samuel de Champlain, in a number of trips from 1609 to 1618, supported by a host of soldiers, builders and labourers, explored the St Lawrence Gulf area of what is now known as New England. Cabeza de Vaca followed a circuitous course through the southeast quadrant of North America on an unplanned journey lasting from 1527 to 1537 (Vaca 1961). His party intended to carry out a brief reconnaissance, but through a series of disasters lost their ships, their commander, their arms, their clothing and their way.

These two narrators are bounded the same way geographically and historically. Behind them, Europe; before them, the utterly unknown. Behind them, land stabilized by representation; before them, an unformed and unsignifying universe. Both for the subjects they are and the world they encounter, the explorers maintain an ideal of stable, rationalized space while occupying a space that is chaotic and mobile. The externalization and control of space the texts seek to propagate goes hand in hand with their attempt to formulate a safely encapsulated subject; cartography seems the ideal method for establishing both.

I select the following passage from *The Voyages of Samuel de Champlain 1604–1618* nearly at random. These descriptions of 'Long Island' appear in two succeeding paragraphs:

> Being distant quarter of a league from the coast, we went to an island called Long Island, lying north-north-east and south-south-west, which

makes an opening into the great Bay Francoise, so named by Sieur de Monts.

This island is six leagues long, and nearly a league broad in some places, in others only a quarter of a league. It is covered with an abundance of wood, such as pines and birch.

(Champlain 1907: 30)

Of these sentences, only one dwells upon sensible, physical characteristics of the island (its trees). There is almost no paragraph in the text that does not exhibit the same qualities: a listing of features with little judgement (beyond whether the land is favourable to commerce and sustenance); little affect; great amounts of instrumental (in both senses) information. The cartographer removes himself from the actual landscape. Though relation there must be for perception to occur, he describes it as much as possible as if he were not there, as if no one is there, as if the island he details exists wholly outside any act of human perception. At the inception of his narrative, Champlain is able to maintain the ideal of an encapsulated, independent space for his subjectivity that will be the hallmark of Cartesian monadism, where the relationship between subject and environment is attenuated, the second term evacuated to a high degree to ensure the uncontaminated primacy of the self.

Part of the function of mapping, it would seem, is to ensure that the relationship between knower and known remains unidirectional. The mapper should be able to 'master' his environment, occupy a secure and superior position in relation to it, without it affecting him in return. This stance of superiority crumbles when the explorers' cartographic aptitude deteriorates. To actually be *in* the surroundings, incapable of separating one's self from them in a larger objective representation, is to be lost. Because Champlain and de Vaca are foreigners, their 'being lost' signifies their real location in the New World. There they could not assume the position of mastery they possessed in their homelands, where their travels carried them (one would suppose) to destinations known in advance across already-ordered spaces.

The environment that the explorers experienced may have had little to do with the fixed space we are accustomed to occupying now. This passage, from his 'Voyage of 1611', may better communicate what the experience of America was like in the beginning. The land he faced itself appeared to have some of the fluid characteristics of this world of water, floating ice, obscuring fog and darkness:

The most self-possessed would have lost all judgement in such a juncture; even the greatest navigator in the world. What alarmed us still more [than the ice] was the short distance we could see, and the fact that the night was coming on, and that we could not make a shift of a quarter of a league without finding a bank or some ice . . .

(Champlain 1907: 199–200)

The land Champlain faces appears chaotic and unstable, moving in its own unpredictable logic. Champlain's vision and his consciousness are increasingly compressed; the land's attributes are magnified until where he *is* seems the whole world. Since he does not know where he is, the environment, rather than being a stable field he moves across, appears to be reorganizing around him. The landscape penetrates the subject – he can no longer maintain his position of cool distance.

For Champlain, whose mission is overall successful, such rendings of the veil of mastery are rare. In spite of his long cohabitation with the Indians, de Vaca remains 'lost' throughout his journey. He feels 'lost' even when the Indians he accompanies are perfectly oriented, because his concept of orientation relies on separating himself from a place, rather than becoming integrally involved with it. Being 'lost' not only describes the subject in space; it describes the subject *as* space. The elevation of the subject over its surrounding space collapses; the minute vacuum assuring their separation disintegrates, likewise decomposing the pure compartmentalization of the subject. 'Being lost' becomes something like a crisis of differentiation, a dysfunction of the logic ensuring ordered space.

De Vaca receives frequent invitations to 'become' an 'Indian'. But the explorers are incapable, ideologically as well as practically, of doing so. Colonization was from the start arranged hierarchically, not as a meeting between equals; the explorers wish to influence and possess the world they meet, but take great pains to be sure that it will not substantially inform them in return. They evacuate the others they meet, keeping their own subject position in the form of the already formulated, complete monad. They want to get *away* from America and its inhabitants, keep it 'outside' themselves, and not change dramatically as a result of their relationship with it.

The European explorers attempted to maintain the environment on the 'outside' in order to preserve their mastery of it. As de Vaca's testimonies relate, the Europeans also needed to maintain the Native Americans as external in order to reinforce their own subjectivity. By attributing inferiority to the Native American peoples and their spatial practices, these texts functioned to concretize individualism and ensure the Native Americans' exclusion from it. Mapping acted to distinguish 'self' from 'other': in early America, cartography was the measure between human and non-human, civilized and savage. Frederick Turner writes, 'Indeed, the primitives' harmonious and precise knowledge of their habitats came in the process of the "Europeanization" of the globe to be the very mark of the primitive itself' (1980: 11). The solid lines that cartography draws between the subject and the land also reinforce the lines drawn between European white subjects and Others. Mapping becomes a technology advancing, and the very hallmark of, a larger cultural order premised on cleanly distinguishing between entities in the natural environment, the psychic environment and, finally, the social environment.

In the last few decades, the Enlightenment subject mapping helped fortify has come under fire – not only in post-structural criticism, but also, Fredric Jameson argues, in its everyday attempt to negotiate external (cultural and physical) space.

Jameson's essay 'Postmodernism, or the cultural logic of late capitalism' (1984) describes postmodernism as a scene in which the matrix of space has abolished categories of time, but space itself has exceeded its traditional organization. He sees postmodernism as an expanding chaos of stimuli unordered by a selective grid of meaning, between whose elements there are no hierarchies, and within which distance and difference are increasingly collapsed. Jameson pairs the new unmanageability of built space with the slipping away of an ordered, obvious direction for intellectual and political practice. (Here he is not so far from the explorers, as in either situation, the whole purpose of representing space is to determine a way to *act*.)

His description of late capitalist space reaches its intentionally dizzying culmination in a description of the Bonaventure Hotel in Los Angeles:

> I am more at a loss when it comes to conveying the thing itself, the experience of space you undergo . . . I am tempted to say that such space makes it impossible for us to use the language of volume or volumes any longer . . . You are in this hyperspace up to your eyes and body . . . It has been obvious, since the very opening of the hotel in 1977, that nobody could ever find any of these stores . . . So I come finally to my principal point here, that this latest mutation in space – postmodern hyperspace – has finally succeeded in transcending the capacities of the individual human body to locate itself, to organize its immediate surroundings perceptually, and cognitively to map its position in a mappable external world.
>
> (Jameson 1984: 82–3)

This passage seems oddly familiar, following upon Champlain and de Vaca. Jameson's essay reproduces on a higher level the anxieties and assumptions concerning space, place and orientation that mark the early explorers' texts. Its thrust is hardly conservative, as Jameson carves out for himself an authorial position both strongly Marxist and productively engaged with postmodern culture. Yet not unlike Champlain, Jameson tends to see the post-modern landscape as a problem, one that needs clearing up. His usual forms of orientation are disabled; he finds no clearly marked or familiar reference points in the way he is accustomed to thinking of them. This space, like that of the explorers, erases the meaningful discriminations of 'here' and 'there', 'before' and 'after', that allow the plotting of a trajectory. Given the way he has framed the 'problem' of the postmodern landscape, Jameson's solution – 'the need for maps' – arrives as a great relief. It is his contention that post-modern space is a 'problem' that I will be investigating: for whom? and why?[3]

The power of Jameson's assertions comes from his supposition that anyone who entered this space would experience it the same way. But, like the Renaissance explorers, he is issuing onto an unfamiliar landscape and seeking to treat it as a familiar one. The implied 'I' here or the frequently used 'we' is not the native entering a transformed world, as he would have us believe, but a foreigner exploring new and unfamiliar terrain. Surely if he visited this site frequently, as a shop girl or maintenance man, or occupied it permanently – if the homeless were allowed entry here – he would gain a working knowledge of it. Like the colonialists, his very ability to encounter this foreign space results from the economic and cultural privilege allowing him to travel around to unfamiliar landscapes – for them, as agents of capital; for him, as its heir. He is a tourist.

Moreover, he has alienated *himself* from this environment in some ways by being an academic Marxist, who wishes to stand outside the edifices of a capitalist world structure, in opposition to it. His method of cultural orientation, classical Marxism, requires that he, like the earlier mappers (both are located by the predicates of science), needs to be at an objective distance from the phenomena he seeks to analyse. The disorientation he attributes to a change in the environment might be charged instead to his navigational method: note that he defines the world as an 'external mappable' world. What we see when we read Jameson's essay is the deterioration of the space established by the explorers and the crisis of the subject associated with it, the monadic 'space-capsule'.

It could be observed that Jameson identifies in the postmodern landscape a *derealization* of space, its plasticity, its tendency to become an infinite semiosis with no resting point. His essay strains to reinstate boundaries that are rapidly becoming far too ephemeral; one of the most representative features of postmodernism, as Jameson describes it, is its erasure of lines that had previously kept separate phenomena and objects apart. Boundaries of all kinds are on the one hand highly emphasized and deemed necessary, but on the other, problematized and metaphorized. In Jameson's postmodernism, all coalesces: the inside becomes indistinguishable from the outside, flatness surpasses depth, the surface melts and takes on a luminous presence separate from the object itself. From the great, concentric loops of the 'sealed membrane' of Edvard Munch's homunculus in *The Scream* (1984: 62–3), to the vacuity of present-day film personalities (1984: 68), to the glossy polyester skin of Duane Hanson's 'dead and flesh-colored simulacra' displacing reality with 'stereoscopic illusion' ('But is this now a terrifying or exhilarating experience?' (1984: 76–7)), to the glass skin of the hotel, like a cop's mirrored sunglasses in its inhuman aggressivity (1984: 82), surfaces and borders are put into a derealizing play reminiscent less of the frontier-bursting transgression of laughter than the out-of-control feeling of a carnival ride. The individual who seeks to negotiate this landscape suffers defeat, anxiety and confusion.

The postmodern subject, like the postmodern landscape it occupies in a relation of mutual reinforcement, has lost its traditional form of closed interiority encapsulated in a boundary. Jameson suggests that 'schizophrenia' aptly describes the state of postmodern subjects in general. While nominally a dysfunction of time, schizophrenia equally presents a dysfunction of space: a failure to adhere to an external reality, to arbitrate the distinction between inside and outside, and to hold the surrounding world together in a meaningful totality. The perception of such a subject would be an unmediated barrage of disordered stimuli whose immediate presence assaults the surfaces of the exposed subject. Moreover, this fragmented subject, barely or only occasionally differentiated from its surrounding world, appears incapable of formulating intentional or oppositional activity. Jameson asks, 'does it [postmodern space] not tend to demobilize us and to surrender us to passivity and helplessness, by systematically obliterating possibilities of action under the impenetrable fog of historical inevitability?' (1984: 86). The immobilization depicted in the preceding question is uncannily similar to the disorientation described by Champlain, yet more important than their superficial similarity is a serious question. How can political activity be imagined, without our simply denying the changes in the cultural complex we currently face?

Jameson's essay distributes the space of the subject in two opposing categories: on one side lies the securely bounded ego of the Enlightenment, a construct allowing instrumental operations but preventing substantial change and seemingly setting subjects into immutable hierarchies; on the other, a fleeting, fractured, postmodern subject no more able to formulate intentions and maintain interests than to maintain one proper unified shape. Such a rigid differentiation of spatial formats represses possibilities for discussing ways spaces, shapes and forms for identity might be substantially altered without being entirely evaporated; for exploring the ways that boundaries might be both maintained and altered to allow political redefinition of the environment.

Like the early explorers, Jameson's ideal of monadic subjectivity is not sustained by his narrative. As with Champlain and de Vaca, the hermetic individual of investigation is shadowed by a less self-assured narrative twin: this subject 'en proces', and how it came to be, is actually the subject matter of Jameson's enquiry. And an interesting subject it is.

What I have always found most intriguing (even exciting) about Jameson's account of postmodern hyperspace is the way Jameson describes space as a medium penetrating him, one that has overcome the very limits of his psychic and physical portals, one that encroaches upon him right up to 'his eyes and body'. The essay 'Cognitive mapping', an early draft of the argument here, betrays similar concerns over the interpenetration of space and body and the connection of body and psyche. There Jameson worries that the 'postmodern body' 'is now exposed to a perceptual barrage of immediacy from which all sheltering layers and intervening mediations have been removed' (1988: 351). This account appeals to me partly because of its obvious eroticism, and partly because it describes so well the experience I undergo traversing all kinds of public spaces – not only postmodern ones, and not only when I am lost. Jameson's comments strike a chord in me: I recognize this state; I enjoy its representation at the same time that I feel ambivalence about the state itself. The crisis of boundaries I undergo in public spaces is, on the one hand, detestable (and for me probably more than Jameson, a sign of the physical, as well as ontological, dangers I face). On the other hand, I experience it as a continually promising phenomenon. It is a sign of a real and vital contact with an outside and an 'other', and an opportunity for a substantial interaction and personal transformation. The apparent surprise that this experience imposes on Jameson, as opposed to its familiarity to me, leads me to posit a gender differential in spatial negotiation. One feature of his spatial anxiety may be the way this space makes his body become conscious to him, an occurrence that is unusual, as he is accustomed – far more than the early explorers, no doubt – to forget the body, to use orienting principles that allow him to erase his physicality. This 'forgetfulness' is, for women in the West, much less available (and a real relief when it does occur). To become conscious of embodiment could only be a positive step for masculinity, as much as such consciousness is also a perpetually wearing aspect of femininity.

A colleague of mine commented concerning this essay that a woman would never be lost as Jameson is: she, because of the ever-present threat of physical attack, is always quite conscious of the position of exits, darkened stairwells and blind corners. Her testimony does not indicate that she is any better 'located', in the teleological sense Jameson intends, than he, but it does suggest that 'orientation' is not a generalizable project. Space can be negotiated on a number of different levels, and for different reasons. Jameson will often refer to his new programme as 'cognitive mapping', apparently assuming that cognitive maps will reintroduce a common ground of perception and understanding. But will a standardization in theorizing spaces exclude, once again, the concerns of subjects who don't fit the model of 'universal' subjectivity? As subjective as the essay on postmodernism may appear, Jameson prefers scientific mapping over older forms of orientation, like the sailor's itinerary, which remain too subjective and ungeneralizable by being tied to specific places and routes and individual intentions (1984: 90).

This text's blind corners derive not from a failure to include all subjects, but rather from the very exclusive/inclusive form it attributes the subject – a form not suitable to describe the dynamics of some non-dominant subjectivities. Mary Ellen Mazey and David Lee, in their comprehensive book *Her Space, Her Place: A Geography of Women* (1983), both define cognitive mapping and demonstrate the disparities that can exist between cognitive maps. From Yi-Fu Tuan (1974: 62), Mazey and Lee take the example of a married couple strolling on a shopping expedition. Though both are (objectively) occupying the same space, the two may not see or hear the same things; their worlds converge only occasionally when the one asks the other to 'admire some golf clubs in the shop window' (Mazey and Lee 1983: 37). The man and the woman may be forming the space around them in such wholly opposed ways, Mazey and Lee contend, that it would be fair to say they are in different spaces.

Let me imagine for a moment the way this landscape might appear to the characters in their traditional couple: for the man, I imagine that the most prominent feature of the landscape would be pathways, along which he projects himself, making his world a space that returns to him a self-image of movement, command, self-assurance and self-satisfaction. For the woman – and, as a woman – I imagine a world structured not by pathways but by obstacles: the people in it may be threats as much as impediments; rather than seeing how to get from 'point A' to 'point B' I often see what is keeping me from getting there.

My hypothetical 'male' perspective may apply only to those men who *are* accustomed to dominating their landscapes – white, youngish, physically able professional-class men; certainly teenage men could not be accused of being ignorant of the attitudes others (especially peers) project towards them, and men in poor urban areas of America cannot afford to be. But I would also argue that men of almost any class or race have the luxury to be far more selective in their environmental attention than women of the same demographic group.

Not only because they are at lesser (perceived) risk of physical attack, but also because of the customary difference of men's and women's labour – even when both are professionals.

Men's work tends to be localized, attached to particular places and time periods. That is, when men are 'off work', they are not working. Women – even those who do not have children – are often responsible for the maintenance of the household's physical environment. They are rarely 'off work' when at home. Men can separate themselves from their environments, live in a space that somebody else creates and maintains, 'tune out', see in the space only what it pleases them to look at.[4] Women, the working class, and people of the Third World create the environment *for* Western men, so they are able to expel it from their consciousness. A woman's consciousness is more immersed in her surroundings, which she – more than a man – is likely to be monitoring for danger or for dust.[5]

There may prove to be, then, different forms of relating to space than that implied by mapping, ones that continue to be practised today by those people who literally cannot afford to separate themselves from the ground: the indigenous, the indigent; until recently, women; and especially, I think, children.[6] Mapping – theoretical or scientific – excludes these subjectively variable perspectives on epistemology, but more importantly, it ignores the variability of subjective structure. Formulating 'subject' as individual with pre-set boundaries, it fails to recognize the very conventionality of the individual boundary that it imposes.

The 'crisis' in subjectivity that Jameson depicts may be largely a crisis only for those subjects who previously were able to establish dominance over their surroundings. The journals of the Renaissance explorers show the hermetic form of individualism to be a historically contingent fiction. Jameson's work suggests, further, that that subject's time has passed.

While Jameson acknowledges the reactionary status of the *modern* subject's stiff differentiation, he does not want the dissipation of subjective boundaries to progress too far. Other post-structuralist critics contend otherwise: the new plasticity of limits in the postmodern era might be precisely the opening that political criticism needs to achieve a radical reformation of subjects. Jameson protests postmodernism's tendency to turn 'everything in our social life . . . to the very structure of the psyche itself . . . "cultural" in some original and as yet untheorized sense' (1984: 87). He does not take this observation to its logical conclusion – that the boundary always was arbitrary and 'cultural'. Hence he disposes the very possibility that has caused post-structuralism to focus on the boundary – that the world can be re-envisioned and revised via human negotiation.

Investigative frameworks like classical Marxism that focus on time position their objects (or subjects) of study as closed entities that maintain continuity as they pass through time; space raises some very difficult questions about the constitution of the object (or subject) itself: where does it begin and end? Where are its boundaries? What differentiates it from other aspects of reality? Taking up space can, therefore, be a truly destabilizing method, as a spatial perspective cannot assume the existence of objects prior to its analysis.

Theoretical mapping, like real Renaissance mapping, shapes subject and environment in a particular way that would exclude some of the very promise of the postmodern – its tendency towards flux and revision; its porousness of division; the fluidity of its boundaries. The inclusive transformations we imagine might require eradicating, radically, the ordering lines of our culture, and our selves.

NOTES

1 Lacan objects to its pretence of self-knowledge and self-transparency (particularly in 'Of the subject of certainty', 1977). Jessica Benjamin (1988) critiques its presumption of independence, the way it represses its inevitable relatedness to others, banishing them to preserve its omnipotence. Feminist critics of science such as Evelyn Fox Keller (1985) and Sandra Harding (1986) have pointed out its participation in a structure of knowledge inseparable from domination, arguing that the empirical subject relates to the world only by objectifying it. Post-colonial critics point out how the imperial subject held the foreign lands and the people in them apart, separate from the 'self', and basically unlike it, or, alternately, incorporated the people and territories they encountered into their own self-image, obliterating their difference. For examples of these perspectives, see JanMohamed (1986) and Spivak (1991).

2 Cultural geography seeks to demonstrate that the organization of space, like any other aspect of the real, is conventional, to some degree arbitrary and indisputably culturally specific. I thank the late J.B. Harley for patiently directing me through this important field of study, still little known in many English departments, and for the trenchant analyses his essays provide.

3 For further discussion of these issues see Jameson (1988, 1991).

4 One of the finest analyses of gender-differentiated modes of spatial existence comes, not surprisingly, from outside the academic community, as well as from a feminist: from columnist Jacqueline Mitchard's series 'The rest of us'. Her article 'Men still fail to pick up on all the pieces' (*Milwaukee Journal* 30 July 1989: G1) provoked much of the thought in the preceding paragraphs.

5 Mazey and Lee provide two fascinating pieces of data that may have a correlation to this divergence in subjective stance. They point out that the enormously profitable genre of romance fiction – supported mainly by women – is practically defined by its depiction of exotic locales (1983: 44). This suggests to me two things: first, that women want to travel, but are compelled to do it vicariously from the safety of their homes (on this supposed safety, however, Mazey and Lee relay the data of a 1978 study showing that 32 per cent of rapes occur in the victim's home (1983: 44)). This dependence on fictions suggests that women have a much harder time escaping mentally, requiring outside support. In relation to the difficulty women have in escaping, Mazey and Lee refer to another study showing that 'women, more than men, rate highest those vacations which would make them carefree, adventurous, daring . . . Though both women and men rate stay-at-home vacations low, men were more enthusiastic about them than women' (1983: 45). This divergence suggests, again, that men more than women are able to divorce themselves from the environment of the home. This data comes from Rubinstein (1980).

6 My research comes on the heels of a furore over the failure of public education to successfully train youths in national and global geography. Children develop naturally a precise and complex knowledge of their 'home' territories, however narrow that 'home' may be. Given that so few children will have an opportunity to intervene (for better or worse) in world politics, is teaching them geography not simply a way of adapting them to the multinational military-industrial complex, without giving them the tools to change it?

4

AS IF THE MIRRORS HAD BLED

Masculine dwelling, masculinist theory and feminist masquerade

Gillian Rose

The transition to a new age in turn necessitates a new perception and a new conception of *space–time*, the *inhabiting of places*, and of *containers*, or *envelopes of identity*.

(Irigaray 1993a: 7)

How, then, are we to try to redefine this language work that would leave space for the feminine? Let us say that every dichotomizing – and at the same time redoubling – break, including the one between enunciation and utterance, has to be disrupted. Nothing is ever to be *posited* that is not also reversed and caught up again in the *supplementarity of this reversal*.

(Irigaray 1985a: 79–80)

[This paper was originally written as a script for a particular kind of performance: a seminar presentation. Performing it produced myself as a speaker and, now, in its essay version, as an author. But, as Butler (1990, 1993) suggests, gender is also produced through performances, and I have argued elsewhere that many perform-ances of academic geography, in their conventionalized ways of talking, writing, behaving, depend on and then reproduce certain masculinities (Rose 1993). However, Butler suggests too that some kinds of performance can destabilize gender. So this paper tries to destabilize the implicitly masculine subjectivity which constructs so many geographical performances by itself performing, performing obviously, performing to excess, performing in different voices. By shifting voices I want to emphasize the constructedness of geographical discourse, of geographical knowledge. The performance is in these shifts between different voices and not in any one of them. I am not trying to advocate an alternative, 'feminine' voice to counter masculinist geography; rather, I am trying to suggest the different critical possibilities offered by a certain mobility between different voices. And the particular object of my mobile critique here is some recent discussions by geographers about real space. So, imagine a cold, analytical, academic voice and read on . . .]

Many geographers have remarked on the prevalence of what they term 'spatial metaphors' in contemporary social and cultural theory. They have noted that subjects are often diagnosed as decentred; that theory travels, knowledges are local, identity is deterritorialized; that nomads, vagabonds and exiles are proliferating; that epistemologies of the margin, the borderland, the diaspora and the closet are being elaborated; that cognitive maps of this postmodern moment are being demanded. Geographers have recognized that such spatial metaphors are widely used in current social and cultural theorizing as a means of articulating the intersection of subjectivity, power and the production of knowledge.[1] The politics of knowledge is understood in terms of the politics of representation, and the politics of representation is interpreted in terms of a geopolitics of location. The production of knowledge, then, is thus always claimed as situated (Haraway 1991). The resultant turn to spatial metaphors in work addressing power/knowledge/identity has led Soja (1987: 289) to suggest that 'space and geography may be displacing the primacy of time and history as the distinctively significant interpretive dimensions of the contemporary period'.

However, many geographers have also been highly critical of the deployment of these metaphors, and most have based their critique on a distinction between the space they argue is assumed in these metaphors and another kind of space they consider to be more truly geographical. In the most detailed critique so far, for example, Smith and Katz (1993) focus on what they describe as the difference between 'metaphorical' space and 'material' space. They begin their discussion by noting that metaphor works by comparing one, unfamiliar meaning system – the 'target domain' – with another, familiar meaning system – the 'source domain' – in order to 'reinscribe the unfamiliar event, experience or social relation as utterly known' (1993: 69). They then offer a brief reading of Althusser and Foucault, chosen for their formative influence on contemporary social and cultural theorizing, and argue that the source domain of the metaphorical spaces in the work of Althusser and Foucault is what Lefebvre described as absolute space: it is transparent, stable and fixed. And Smith and Katz suggest that 'it is precisely this apparent familiarity of space, the givenness of space, its fixity and inertness, that makes a spatial grammar so fertile for metaphoric appropriation' (1993: 68). They then go on to argue that metaphors which refer to absolute space function to freeze what are in fact dynamic social processes; thus they accuse Foucault's 'pervasive substitution of spatial metaphor for social structure, institution and situation' of continuing 'to elide the agency through which social space and social relations are produced' (1993: 73). In contrast, Smith and Katz describe material space, which they also term 'geographical space' and 'social space' (1993: 73, 80), in terms of its 'multiple qualities, types, properties and attributes . . . and its relationality' (1993: 80). And they conclude that if spatial metaphors are to be part of a radical critical project, it is this material or geographical or social space which must be their source domain, because its dynamism renders socio-spatial structures amenable to change; in contrast, spatial metaphors which refer to absolute space are regressive because absolute space serves to freeze and thus to sanction the socio-spatial or theoretical status quo.

Smith and Katz's insistence on the need to interrogate critically the specific qualities of the space to which social and cultural theorists refer is without doubt correct, and many radical geographers have made similar arguments (Bondi 1993; Keith and Pile 1993; Massey 1993; Pratt 1992; Reichert 1992). However, Smith and Katz legitimate their argument by claiming that the material, geographical, social space they advocate is real. Material, geographical, social space is actually real space. And they are not alone in making this claim: their essay is but the most full elaboration to date of a distinction which many geographers are currently making between what they argue is a real space and what in contradistinction they define as some kind of non-real space. For Harvey (1989), the distinction is between 'real' and 'metaphorical' space or between reality and image; Hanson (1992: 573) differentiates between what she terms 'geographic space' and 'cultural space'; Daniels and Cosgrove (1993: 57) separate metaphorical from actual space; Bondi (1992) distinguishes between 'real geographies' and 'symbolic' ones; Agnew (1993: 261) differentiates representations and metaphors of space from concrete particulars; and Lagopoulos (1993: 264, 265) makes distinctions between cognitive and geographical space and also between signified space and real space. Implicit in these distinctions (and often explicit also) is a hierarchization: real space is understood as a more accurate description of causal processes, and is therefore more important for geographers to study. For all of these geographers, then, there is a real space to which it is appropriate for metaphors to refer, and a non-real space which it is not.

In what follows, this distinction between real and non-real space is characterized as a performance of power, and of masculinist power in particular. It will be shown that the distinction between real and non-real space is constructed in terms which are also gendered, and that this hierarchical engendering of spaces is naturalized by the claim that only one of those spaces is real. The distinction between real and non-real space thus requires a feminist deconstruction.

This deconstruction is not difficult to develop, however. The distinction between real and non-real space does not hold even among its advocates, since the distinction is not signified consistently in their discussions. The distinction produces instabilities which undermine the distinction itself. To understand this failure, Butler's (1990, 1993) discussion of performances of hegemonic power are useful. For Butler, discursive power performs its productive effect through its reiteration of naturalizing norms; it enacts what it names. The reiteration of the real by geographers can be seen in the multiplication of terms used to describe that space: real, material, experienced, concrete, social, actual, geographical. The non-real proliferates too among geographers: imagined, symbolic, metaphorical, imaged, cultural. These are the reiterations which for Butler constitute the performance of normative power. However, Butler goes on to argue that norms must be performed repeatedly because their constituting citation of the subject and non-subject is never guaranteed. Reiterations thus also constitute the failure of discursive power; their very repetition assumes a failure in their effect, and it also produces gaps and fissures in discourse which can be resources for critique. In the context of normative geographical performances, the latter possibility is particularly pertinent because the repeated terms used to define real space are not equivalent. The 'real', 'experienced', 'concrete', 'social', 'actual', 'geographical' are not the same things, and nor is any one consistently aligned with a non-real opposite. The excess of these terms – that is, their non-equivalence and multiplication – suggests that the distinction between real and non-real space is not a self-evident one. Indeed, it suggests that the distinction cannot be maintained with either clarity or firmness in these accounts.

If the distinction between the real and non-real is unstable, then, so too are the definitions of each. This is particularly evident in relation to the tropes of fixity and dynamism which are central to the constitution of the 'real' and the 'non-real'. I have already noted that for many geographers, what signifies real space is its dynamism, its processual creation, its construction. In contrast, what signifies non-real space is its fixity, immobility, stasis. However, assertions of the reality of real space consistently also invoke images of fixity. Thus Harvey (1989: 187) describes real space in terms of 'its invariant elements and relations', Smith (1984: xiv) wants to analyse its 'concrete process and pattern', and Soja and Hooper (1993: 195) also describe real space as 'concrete'.[2] Real space then becomes the 'very solid basis' from which Harvey (1989: 112), for example, can interpret the world. And if the real is solid, concrete and invariant, then the non-real, all that the real is not, must be signified not in terms of fixity but in terms of fluidity. Thus Smith (1984: xiv) refers to Nature as 'sodden with metaphor', Smith and Katz (1993: 69) describe non-real space as 'fertile', and in Harvey's (1989) hysterical account, non-real spaces become a maelstrom in which all that is solid apparently melts: his imaged spaces are ephemeral, superficial, specular, eclectic, fragmented, chaotic, depthless, schizoid, fascistic, fetishistic, a fluid towards which one is seduced by veiled, titillating gyrations and in which one then wallows or, even worse, is swamped. In this account, the non-real signifies not fixity but total dissolution.

Here it is impossible not to remark on the rhetorical encoding of two different kinds of space as two different kinds of sexes. As with so many binary accounts of difference, the difference between these real and non-real spaces is constructed through the terms of sexual difference. The real is simultaneously concrete and dynamic, yet both these qualities signify the masculine; the non-real is simultaneously fluid and imprisoning, but always engendered as feminine. Material real space could thus be re-described as the effect of masculinist power, its very materiality also its particular masculinity; but non-real space is also the effect of masculinist power, its lack of reality the sign of its feminization. The instabilities between and within these efforts to define real and non-real space are symptomatic of, indeed are constitutive of, a compulsive fixing of sexual difference. They are so chronic as to displace the distinction itself to reveal what Derrida (1979: 293) would describe as 'the common ground', or the *différance*, on which the distinction depends. What the instabilities of this distinction suggest is that the distinction is a dualism which reiterates the constitutive relation between the masculine same and the feminine other. Through trying to fix difference, they fix the same. It is difficult not to agree, then, with Butler (1993: 2), when she argues that materiality (by which she means that which is naturalized as real) should be 'rethought as the effect of power, as power's most productive effect'. The reiteration of the distinction between real and non-real space serves to naturalize certain, masculinist visions of real space and real geography, and to maintain other modes of critique as, to quote Christopherson (1989), 'outside the project'. It is an act of exclusion.

This reading of the repetitions and reversals in the terms which are intended by some geographers to secure the distinction between real and non-real space suggests that the distinction, and therefore also its exclusions, cannot be made to hold. The work undertaken to stabilize the distinction also works to collapse it. Indeed, its collapse suggests that there are not two objects at stake in this discussion – real space and non-real space – at all, but just one – space – which is split into two. Space is split into two and becomes, in the words of Bhabha (1990: 76), 'at once an "other" and yet entirely knowable and visible'. And for Morris as for Bhabha, this splitting indicates an ambivalence towards its object: Morris (1992: 262) describes ambivalence as 'a way of struggling with the problem posed by an impossible object – at once benevolent and hostile – by splitting it in two'. The conclusion must then be that for those geographers concerned to maintain the distinction between real and non-real space, space itself is their impossible object.

[A pause, and then another voice, enthusing, a young lecturer perhaps, eager to interest and to be interesting . . .]

Well, it looks like if I want to say anything more about this I've got to begin again, somewhere else, beyond geography! What I want to do in this chapter is to challenge the distinction a lot of geographers are making right now between real and non-real space by talking about, or talking with, the work of Luce Irigaray. And I want to begin by explaining why I love Luce . . . (sadly that joke isn't my own – it's from Chris Holmlund, 1989).

Now, Luce Irigaray is not a geographer – she's trained in linguistics and psychoanalysis, and she also talks about philosophy in detail. Even so, space is fundamental to the way she thinks about subjectivity, language and power. Spatialized notions – like extension, gap, interval, container, envelope, horizon, verticality, threshold, geometry, dwelling, elsewhere – are all key terms in her arguments about phallocentric structures of meaning and phallocratic social institutions. She uses these spatial terms carefully, although not always consistently, throughout her work, and so I'd argue that she is a kind of theorist of the spatial – although I think part of her project is to redefine what we might understand by both 'theorist' and 'spatial'.

So her use of spatial terms doesn't mean that she's some kind of geographer by another name. Quite the opposite – and this is one reason why I like her work so much – her discussion of space occupies an entirely different register from that of geography. And this is most obvious to me in the way she writes. Sometimes her work reads like that of an incisive, clinically dispassionate critic, deconstructing texts with a cold, relentless rigour. More often though she is emotional, elliptical and allusive. Her critiques of Western metaphysics sometimes sound like sad letters from a grieving lover, or maybe mother, and at other times like a tirade from a well-informed biology mistress. It was this mobility of voice got me thinking about the critical possibilities of performing this piece, of working with the uncertainties of performance that Butler (1990, 1993) talks about, trying to parody and masquerade in different voices.[3]

The richness and mobility of Irigaray's writing voices have the effect of making the monotone of most academic writing sound less like an analysis and more like a symptom. And of course Irigaray argues that is precisely what dominant modes of academic writing are, a symptom. In the flat monotony of the philosophical text, the psychoanalytic text and the geographical text – and I just love those kind of generalizations she makes even though I know I shouldn't, they're so ridiculously, wonderfully sweeping – she hears the repetitive and compulsive iteration of phallocentrism. So she writes to disturb that monotone, that monotone that's also a monologue, and she does that in part by her diverse writing strategies which are a kind of teasing, 'hav[ing] a fling with the philosophers' (Irigaray 1985a: 150).

She also disturbs that monotone by establishing conversations *within* her texts – she often addresses a series of questions at quotations from other authors. And she also establishes conversations *with* her texts. Irigaray invites a conversation between her writing and myself or you as her reader – and a major part of Irigaray's project is to enable a dialogue between her and me as women, to enable a dialogue among women. When she talks about reading, she says:

> The only response one can make to the question of the meaning of the text is: read, perceive, experience . . . *Who are you?* is probably the most relevant question to ask of a text, as long as one isn't requesting a kind of identity card or autobiographical anecdote. The answer would be: *how about you?* Can we find common ground? talk? love? create something together? What is there around us and between us that allows this?
>
> (Irigaray 1993b: 178)

Irigaray insists that I remain distinct from her – she wants to make a 'between' between us, an around, a space, in order to initiate a dialogue. She gives me an invitation to speech through this assertion of a kind of connective space between us. And – being obsessed with geography, feeling that it matters so much and that much of geography is not so much doing it wrong as missing the point – what I want to talk with her about is precisely this notion of a 'between' and 'around' her and me. That 'between' and 'around' is a space I find enabling, to me it's real, because it invites me to perform, to respond, to try to produce a self in relation to her. But why does Irigaray spatialize the possibility of dialogue?

This is a question about how space itself is bound into subjectivity and inter-subjectivity, and Irigaray has prompted a risky response from me. She says that 'no narrative, no commentary on a narrative, are enough to produce a change in discourse . . . unless they go beyond the utterance into the creation of new forms' (Irigaray 1993b: 177). So here are some of the questions about some geographers' efforts to distinguish between real and non-real space provoked by conversing with Irigaray.

[And here a third voice seems necessary, inflected by Irigaray's intonations . . .]

. . . what if the insistence that there is a real space, that there is a real which is neither imagined nor symbolic, a real which is quite untainted by the imaginary or the symbolic, is not a statement of plain commonsense fact at all, but a hope, a desire? What if this real, this claim that there is a real space, itself depends on desire, is itself an imagined fantasy? A desire for something safe, something certain, something real? A fantasy too of something all-enveloping, something everywhere, unavoidable, unfailingly supportive: space? In which all things could be charted, positions plotted, dwellings built and inhabited? And was itself a dwelling, secure? What if that space, that real space, was a dream, an old dream, a most basic dream? Would this be space dreamt as territory? Would this be the plenitudinous space which geography assumes, in which everything has its space and no space can be occupied by two objects simultaneously? Might this be the actual space of the real world which can be fully analysed scale by scale, from a satellite maybe, or a hotel tower, each scale with its own real structures, scale after scale nesting one within the other in infinite regress, the latest addition the scale of the body, in a series which hopes to constitute a totality (Smith 1993)? Whose desire, whose space would this be? Who would it constitute? Who would dwell within it, and how?

[Well, that's rather an embarrassing passage measured by the conventions of academic geography, so perhaps I'd better return quickly to the certainties of a more academic assertive voice . . .]

1 Irigaray assumes that space is a medium through which the imaginary relation between self and other is performed.
2 Irigaray assumes that certain formulations of space enable the production of only certain relationships between self and other.
3 Irigaray assumes that the master subject constitutes himself through the performance of a particular space. For example, she suggests that 'it is rather by distance and separation that he will affirm his self-identity' (Irigaray 1985b: 166) and that 'if he arrived at the limits of known spatiality he would lose his favourite game, the game of mastering her' (Irigaray 1993b: 42)
4 Irigaray assumes that the critical task is therefore to subvert the space of the master subject and to remedy the absence of 'the missing categories of her space-time' (quoted in Whitford 1991: 159).

[Order restored, an academic voice returns, but becomes increasingly baffled . . .]

A key term in understanding Irigaray's conceptualization of subjectivity and space is 'imaginary'. As Grosz (1989) suggests, Irigaray's use of the term 'imaginary' bears some relation to its use in Lacanian psychoanalysis. For Lacan, the 'imaginary' refers to the dyadic relation between self and other, or to what Ragland-Sullivan (1992: 174) describes as 'the domain of transference relations', of which the pre-Oedipal relation between mother and child is paradigmatic. For Irigaray too, the imaginary is central to modes of inter-subjectivity. This imaginary relationship is later structured through the cultural constructions of the symbolic. To the extent that she agrees with Lacan, then, Irigaray's notion of the imaginary might be represented as in Figure 4.1.

imaginary   ⟶   symbolic   ⟶   social

'If we could make the foundations of the social order shift, then everything will shift.' (Irigaray 1991: 47)

Figure 4.1

However, there is other evidence in her work to suggest that, agreeing with commentators like Brennan (1991) and Elliott (1992) who insist on a difference between the social and the psychic, the relationship between the social and the imaginary for Irigaray might better be shown as in Figure 4.2.

social   ⟶   symbolic/imaginary

' . . . the absence of male doctors [from a conference on women's mental health] is, in and of itself, one explanation of madness in women: their words are not heard.' (Irigaray 1993b: 10)

Figure 4.2

But then again, something like Figure 4.3 also represents her arguments.

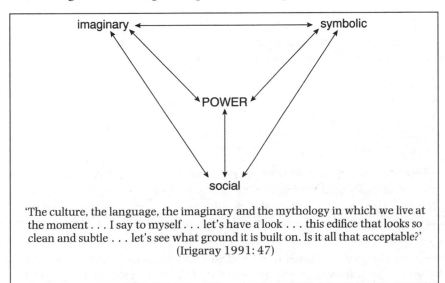

'The culture, the language, the imaginary and the mythology in which we live at the moment . . . I say to myself . . . let's have a look . . . this edifice that looks so clean and subtle . . . let's see what ground it is built on. Is it all that acceptable?' (Irigaray 1991: 47)

Figure 4.3

And if it is not at all clear how Irigaray theorizes the imaginary in relation to its Lacanian formulation, neither does she differentiate in her work between the quite incompatible connotations carried by the concept of 'the imaginary' from its diverse deployments by Lacan, Bachelard, Sartre (Whitford 1986) and by Le Doeuff (Morris 1988). Thus it is impossible to convert Irigaray into anything like the example in Figure 4.4.

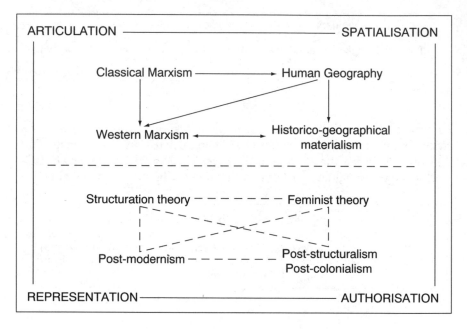

Figure 4.4 Marxism and post-Marxism (Gregory 1994: 101)

How *beautifully* reassuring, what a, literally, incredibly clean edifice, that diagram is. Although perhaps not a very subtle one.

[And here, if I wanted to continue in this particular voice, I'd turn away with a certain amount of relief from the confusions, sorry, the complexities of Irigaray's work and, with a certain amount of relish, turn towards the task of explicating exactly how Derek's, sorry Gregory's work isn't quite complex enough . . . but I'd rather return to being intrigued and enthused by Irigaray . . .]

Irigaray absolutely refuses to define her conceptual apparatus, or to maintain the theoretical stability of any of the terms she uses in her project, or to systematize the relations between them. She refuses to make a clear distinction between the imaginary, the symbolic, fantasy and the social, for example. Addressing psychoanalysts, she says:

> Your fantasies lay down the law. The symbolic, which you impose as a universal innocent of any empirical or historical contingency, is your imaginary transformed into an order, into the social.

> (Irigaray 1991: 94)

Indeed, Irigaray is hostile to the very notion of 'concept' itself. And here I have to talk with her terms . . . they all come rushing in . . . enunciation, imaginary, auto-affection, property . . . I can't assimilate her work; just as she insists on my voice in this dialogue, so she insists on her own. So: she argues that the idea of 'concept' is part of 'a system of meaning which serves the auto-affection of the (masculine) subject' (Irigaray 1985a: 122–3). She argues that the 'concept' is part of the male – well, the male imaginary. For her, 'concept' is an authoritative definition of meaning, closed and proper; it can be traded like a commodity among men in the academic marketplace, she argues, constituting them in the act of exchange as men of theory (Irigaray 1985a: 158), or placed at unitary points in orderly conceptual diagrams. For her, the task is to 'disengage ourselves, *alive*, from their concepts' (Irigaray 1985a: 212). Irigaray resists what she sees as the closure of the conceptual by insisting on the dialogic nature of her texts – her diverse modes of engaging with particular texts, her engagement with her reader – so her work is itself imaginary, explicitly relational. And she argues that this means it cannot be transposed into another context, cannot be appropriated as property into circuits of theoretical exchange or expropriated into diagrams of proper conceptual relations (Irigaray 1993b: 177–8).

I want to say more about Irigaray's notion of the imaginary, but because of her hostility to proper definitions it's easier to talk about what Irigaray sees it is *not* than what it is. Irigaray uses the term to disrupt a number of key conceptual distinctions which underpin dominant, phallocentric modes of thinking.

For example, Irigaray uses the term 'imaginary' to refuse the distinction between concept and practice. This is very obvious in her critique of psychoanalysis. Addressing 'gentlemen, psychoanalysts', she condemns their professional, institutional rejection of anyone 'who questions the history, culture or politics in which psychoanalysis is inscribed', and suggests that this rejection occurs in 'the name of a father of psychoanalysis to whose unconscious any unconscious should be made to conform' (Irigaray 1991: 80, 81). This conformity is also a theoretical move; Irigaray argues that psychoanalytic theory exemplifies phallocentric discourse in which only men can accede to subjectivity because only masculine subjects are acknowledged by that discourse and can thus be symbolically constituted through it; as she notes, for psychoanalysis; 'a man minus the possibility of (re)presenting oneself as a man = a normal woman' (Irigaray 1985b: 27). This intersection of conceptual framework and institutional practice she argues is an effect of the imaginary. The imaginary is at once symbolic and social.

Her use of the term 'imaginary' also displaces the distinction between the individual and the broader social structures in which they are embedded. Thus psychoanalysis and psychoanalysts are simultaneously the object of her critique. And she writes about women's health, for example, and queries:

[women] are often slightly unwell? Maybe. How could this be otherwise when there's no space for a woman's self-affirmation as *I* [*je*], but when, on the contrary, we must continuously support the assertions of others:

in discourse, in images, in actions, and particularly in the commercial use of the self.

(Irigaray 1993c: 101)

This example also indicates the way in which her use of the imaginary displaces the distinction between mind and body, between culture and nature. Bodies are inscribed with the marks of phallocentrism, and phallocentrism itself is structured by interpretations of the bodily. Irigaray (1985a, 1985b) suggests that the male imaginary has a certain topography, of one-ness, verticality, solidity, and that this is bound into male genital 'morphology'. 'Morphology' is Irigaray's term for the symbolic significance of anatomy, a term which emphasizes the cultural encoding of the bodily. The imaginary is thus a 'corporeal imaginary' (Whitford 1991: 59).

Most importantly for what I'm trying to argue here, Irigaray's understanding of the imaginary is also a way of refusing to render fantasy in opposition to 'reality'. She turns again and again, for example, to the significance of Greek myths in order to interpret contemporary Western issues like the use of technology, or environmental pollution, or mental health. Irigaray is also concerned to challenge the fantasies which underpin the apparently objective theoretical frameworks of philosophy and psychoanalysis. These fantasied modes of social and theoretical organization are also the imaginary.

The imaginary, then, could be described as a series of refusals by Irigaray, a series of refusals of dichotomies. The imaginary refuses to distinguish between the social and the symbolic, or the real and the imagined, or the real and the textual, or between the bodily and the cultural, or between agency and structure. And it is this imaginary that I want to think space through. I want to argue that the imaginary is itself spatialized and enacts spatialities; it performs the relation between subjects through spatialized relations. It enacts and produces particular subjectivities through particular spatialities. And in the case of what Irigaray calls the male imaginary, the imaginary performs subjects and non-subjects, space and non-space, same and other.

[I now want to stage an exchange of comments between Irigaray and myself, a dialogue, on the subject of the geographical imaginary which desires a difference between real and non-real space. I cite her often allusive comments on the spatialities of dominant, masculinist modes of knowledge, and then see what interpretations of geographers' efforts to maintain a distinction between real and imagined space might follow . . .]

Irigaray argues that there is a topography to the performance of the male imaginary:

> the *scenography* that makes representation feasible, representation as defined in philosophy, that is, the architectonics of its theatre, its framing in space–time, its geometric organization, its props, its actors, their respective positions, their dialogues, indeed their tragic relations, without overlooking the *mirror*, most often hidden, that allows the logos, the subject, to reduplicate itself, to reflect itself by itself. All these

are interventions on the scene; they ensure its coherence so long as they remain uninterpreted.

(Irigaray 1985a: 75)

Mirrors. Flat mirrors, endlessly reflecting the anxious subject back to himself as he spectates, speculates. Reflecting the same back to himself. The mirror produces the illusion of he who imagines he stands in front of it: the master(ful) subject. The distance from the mirror, its invitation to his gaze, constitutes this subject. This subject is the effect of speculation.

> For relations among subjects have always had recourse, explicitly or more often implicitly, to the flat mirror, that is, to what privileges the relation of man to his fellow man . . . What effects of linear projection, of circular turning back onto the self-(as the) same, what eruptions in signifying-points of identity has it entailed? What 'subject' has ever found in it, finally, its due?

(Irigaray 1985a: 154)

And the mirrors are frozen. Rigid like ice. Are mirrors 'the net you draw round your catch, the ice in which you store your property, the mirrors where you conserve and freeze your desires?' (Irigaray 1992: 18). Solidified in their repetitive reflection of the same, a solidity of morphological tumescence and of death. And mirrors can be walls. They cluster together, overlap, build a 'palace of mirrors' (Irigaray 1985b: 137), provide 'solid walls of principle' (Irigaray 1985a: 106). They give form, they turn ideas into structures, edifices, they produce 'the absolute power of form' (Irigaray 1985a: 110), the solidity of concepts, boundaries and order. All this is part of 'a complicity of long standing between rationality and a mechanics of solids alone' (Irigaray 1985a: 107).

> From this encircling projective machinery, no reality escapes unscathed. Alive. Every 'body' is transformed by it . . . What is disturbing is that of these fantasies he makes laws, going so far as to confuse them with science – which no reality resists. The whole is already circumscribed and determined in and by his discourse . . . Nothing escapes the circularity of this law.

(Irigaray 1985a: 88)

I respond to Irigaray . . . And the desire for the solid produces too among geographers a desire for rules and regulations which might stabilize meaning. For if they are forced to acknowledge metaphor since 'any project to abolish metaphor is not only doomed to failure but is, literally, absurd', then they must also demand that 'it is necessary to devise more explicit translation rules . . . between material and metaphorical space' (Smith and Katz 1993: 68). Indeed, to describe these spatial images as metaphors at all assumes an economy of substitutable solid objects and reiterates 'the privilege granted to metaphor (a quasi solid) over metonomy (which is much more closely related to fluids)' (Irigaray 1985a: 110). Real space then contains a certain fixity, concreteness, solidity.

Irigaray also says . . . and in this corporeal imaginary, the body itself is imagined through mirrors. Mirrors build a body, a carapace, a hard body, enclosed. The mirrors envelop the master subject. 'Proprietor, your skin is hard. A body becomes a prison when it contracts into a whole . . . When a line is drawn around it, its territory mapped out' (Irigaray 1992: 17). And the line creates an inside and an outside.

> But now everything has moved inside the house the subject has made, or is. And whether the scene seems set inside, or outside, whether in his room or in his study, sometimes enjoying a fire fancied to be burning in baroque curls of smoke or else gazing out through the/his window at the still in(de)finite space of the universe, the action is always inside his house, his mind.
>
> (Irigaray 1985b: 212–13)

And in occupying 'the site of sight' in this way (Irigaray 1985b: 95), the subject himself is invisible to himself. Looking out from his palace of mirrors, from his study constituted as an invisible place, the site of sight, the subject sees the world, his world.

And I can continue . . . Rules, rulers, ruled. Metaphorical space, in its proliferation, must be brought to order; the fear of metaphors 'out of control' requires a demand for control (Smith 1993: 98). The 'actual spatial source of such metaphors as domain, field, region' must be stated (Smith and Katz 1993: 73), regulations producing regulators, controlling controlled producers of meaning. They might risk going through the looking-glass like Alice, but they guarantee their return journey (Smith and Katz 1993: 74), not wanting to linger on 'the other side' (Irigaray 1985a: 9–22). And that other?

Irigaray says . . . The very first dependence, the very first relationship, is with the mother, 'this bodiliness shared with the mother, which as yet has no movement of its own, has yet to divide up time or space, has in point of fact no way of measuring the container or the surrounding world or the content or the relations among all these' (Irigaray 1985b: 161). The first relationship is before time or space. But, 'the everything that he once received in his mother's womb: life, home . . . food, air, warmth, movement, etc. This everything is displaced . . . because there is no way to place it in its space, its time, and the exile from both' (Irigaray 1993b: 16). The exile from the mother occurs at birth; and when the mother is absent; through the castration complex; through the symbolic murder of the mother. It has profound consequences. The exile produces a feminine other:

> A woman serves (only) as the *projective* map for the purposes of guaranteeing the totality of the system – the excess factor of its 'greater than all'; she serves as a *geometric prop* for evaluating the 'all' of the extension of each of its 'concepts', including those that are still undetermined, serves as fixed and congealed *intervals* between their definitions in 'language,' and as the possibility of *establishing individual relationships* among these concepts.
>
> (Irigaray 1985a: 108)

She is the 'silent plasticity' which can ensure 'the serene contemplation of empire' (Irigaray 1985b: 142, 136). She is the other of his language – blanks, excess, matter, a matrix – necessary to his meaning but silenced within it.

> If there is no more 'earth' to press down/repress, to work, to represent, but also and always to desire (for one's own), no opaque matter in which theory does not know herself, then what pedestal remains for the existence of the 'subject'?
>
> (Irigaray 1985b: 133)

This is the matter, the nature, the raw material of speculation, on which theories and concepts depend. This is space, 'this sea where he is . . . an *extended corporeal thing*. Probably immense . . . an extension at the "I"'s disposal for analytical investigations, scientific projections, the regulated exercise of the imaginary, the utilitarian practice of technique' (Irigaray 1985b: 185–6). This is woman as place, 'never here and now because she is that everywhere elsewhere from whence the "subject" continues to draw his reserves, his re-sources, yet unable to recognize them/her' (Irigaray 1991: 53). The loss of security, loss of self, disintegration threatened by the exile from the mother is symbolized through the figuration of the (m)other as the symbol of the death which this loss invokes. She, matter, becomes the location of his fear of death and dissolution. The mother functions as a place for the sublimation of death (Irigaray 1985b: 54–5). Her deadliness is her otherness; 'in this proliferating desire of the same, death will be the only representative of an outside, of a heterogeneity, of an other: woman will assume the function of presenting death (of sex/organ), castration' (Irigaray 1985b: 27). The death drive is displaced onto the figure of mother as place, matter, matrix. This is the uncanny place, the feminine place to which heroes are drawn but from which they must also escape.

So I can comment . . . that (space) which is non-real is condemned by geographers as dead, deathly, frozen, icy, calcified, as feminine; and it is condemned, expelled into the realm of the non-real. But whose death does it threaten, whose death is entailed, since this non-real space is expelled because it resists analysis: spatial metaphors 'work precisely by reinforcing the deadness of space and therefore denying us the spatial concepts appropriate to analysing the world' (Smith 1990: 169). It is the inability to analyse which produces death, it produces only the casualties of the battlefield (Smith and Katz 1993: 75, 77), it would be 'fatal' to get the analysis wrong (Merrifield 1993: 526). The non-real threatens the end of analysis, the end of the analyst. So deadly it must be non-real.

Irigaray once more . . . Belief in solids, in solid objects, in the material as simply there, is thus a guarantee of the subject's own subject-hood. A belief in the real is a belief in the self because it is a belief in a death displaced elsewhere, beyond the territory, beyond the protective, reflective tain of the mirror.

> You don't believe it? Because you need/want to believe in 'objects' that are already solidly determined. That is, again, in yourself(-selves),

accepting the silent work of death as a condition of remaining indefectibly 'subject'.

(Irigaray 1985a: 115)

And in return I can say . . . So real space, not being the deadly non-real, must be dynamic, creative, fecund, concretely so. The production of space is lauded, 'the process of producing' explored (Merrifield 1993: 523), its social construction analysed (its construction, its building, its construction as a dwelling) (Bondi 1993; Smith and Katz 1993). It must be the location for 'actual daily life' (Merrifield 1993: 525), it must have a 'lifely tension' (Reichert 1992: 90). A space produced, requiring production, by the subject, producing the subject. A really material, geographical space, a really material geographer. Really real.

Irigaray suggests . . . that his 'vectorizations of space' are the means of suturing the unacknowledged wound of the break with the mother, a re-membering that is also forgetting, dis-membering (Irigaray 1993b: 34–5). 'She can only be encountered piece by piece, step by step . . . if he arrived at the limits of known spatiality he would lose his favourite game, the game of mastering her' (Irigaray 1993b: 42). The fear of the mother mobilizes a particular mode of spatiality, a space of territory, position, containment, distancing. He wants 'to master her, to reduce her little by little to nothing, by constructing for himself all kinds of new enclosures, new homes, new houses, directions, dimensions, foods, in order to break the bond with her' (Irigaray 1993a: 34). And this is a process of dwelling.

> *To inhabit* is the fundamental trait of man's being. Even if this trait remains unconscious, unfulfilled, especially in its ethical dimension, man is forever searching for, building, creating homes for himself everywhere: caves, huts, women, cities, language, concepts, theory, and so on.
>
> (Irigaray 1993a: 141)

And she is contained, enclosed, by these efforts. She is imagined as contained, imagined as having a spatiality of impermeable borders, imagined as having 'the solidity of land' (Irigaray 1991: 64), as being the hard surface of the mirror:

> What 'other' has been reduced by [the mirror] to the hard-to-represent function of the negative? A function enveloped in that glass – and also in its void of reflections – where the historical development of discourse has been projected and assured.
>
> (Irigaray 1985a: 154)

The surface of the mirror is feminized, woman is the flat surface at which the subject gazes, sees himself, gazes that distance. There is 'a play to achieve mastery through an organized set of signifiers that surround, besiege, cleave, out-circle, and out-flank the dangerous, the embracing, the aggressive mother/body' (Irigaray 1985b: 37). And she is enveloped; her envelope consists of clothes, jewels, cosmetics, the home. 'You close me up in house and family. Final, fixed walls . . . Have I ever experienced a skin other than the one which you wanted me to dwell within' (Irigaray 1992: 25, 49). There is no spatiality which would allow the mother to become a subject.

> You grant me space, you grant me my space. But in so doing you have always already taken me away from my expanding place. What you intend for me is the place which is appropriate for the need you have of me. What you reveal to me is the place where you have positioned me, so that I remain available for your needs.
>
> (Irigaray 1992: 47)

Envelopes are another solid then; they depend on a certain kind of space to constitute the masculine subject and his feminine (m)other.

And I repeat . . . so non-real space is static, passive, immobile.

Irigaray continues . . . these feminine envelopes always contain the threat of the fluid body matter contained, though. They contain it to limit it, and they also contain it by carrying it. Inside the envelope blood and milk flow.

> There almost nothing happens except the (re)production of the child. And the flow of some shameful liquid. Horrible to see: bloody. *Fluid* has to remain that secret *remainder*, of the one. Blood, but also milk, sperm, lymph, saliva, spit, tears, humours, gas, waves, airs, fire . . . light. All threaten to deform, propagate, evaporate, consume him, to flow out of him and into another who cannot easily be held on to. The 'subject' identifies himself with/in an almost material consistency that finds everything flowing abhorrent . . . Every body of water becomes a mirror, every sea, ice.
>
> (Irigaray 1985b: 237)

And I want to say . . . hence the horror of imaged, non-real space as fluid, 'that opaque world of supposedly unfathomable differences' (Harvey 1993: 5). The horror of the ephemeral, specular, eclectic, fragmented, chaotic, schizoid, fascistic, fetishistic, fluids, veiling, titillating, enveloping, wallowing, swamping. Fluid and fusing. Hence the fear of non-real space is the fear of 'fusion' between real and non-real space (Smith and Katz 1993: 68), the fear of fusion itself. Horrible, threatening. Surely therefore unreal, unimportant, non-real. The object of desire itself, for geographers, would be the transformation of fluid to solid (cf. Irigaray 1985a: 113)?

For Irigaray it is obvious that 'woman, in this sexual imaginary, is only a more or less obliging prop for the enactment of man's fantasies' (Irigaray 1985a: 25). And she is a spatialized prop: matter, earth, nature, interval, ground, envelope, container. And while he fantasizes a universe from the implosion that is his study, she is both infinite matter and an envelope or mirror, always carrying his meaning.

And for me finally to try and describe the male geographical imaginary, it performs . . .

. . . *an infinite sea of matter, imagined from an enclosed and frozen place. A study requiring an immense scene. A doubled space, its subject constituted by extension and intension, (self)contained and expansive. Only the study and the extended space real.*

. . . *the (m)other both extensive matter and enclosed container. Matrix and envelope. A hole and holed up. Imprisoned and exiled, 'unmarked primary matter on the one hand and the signs or emblems in which he cloaks her on the other' (Irigaray 1993b: 115). The spaces of her surfaces and her fluidity unreal.*

. . . *a desire for solidity but with fluids always threatening.*

. . . *space, then, as an impossible object. The (m)other as both petrifying and fluid, the same as both concrete and processual.*

Yet Irigaray insists . . . the distancing of the mother produces another supplement which undermines the certainty of phallocentrism. The very process of representation contains its own uncertainty: 'he believes that, when he sent her far off like this, she will come back the same, whereas she returns to the other in the same (*le même*). This difference undermines the truth of his language' (Irigaray 1993b: 35). There will always be an elsewhere, beyond the contradictory spatiality of the same and his other. Another way of thinking space, another way of relating to others.

So I now feel I can say . . . in this frozen space, this waste, rigid with its repetitious reflections, its serial thinking, in this palace of mirrors, this sepulchre, the tomb of the corpse of the clanking, glittering carapace of the 'subject', occupying this performed space are others, striating this performed space is another possible space. Some might be parodying their surface appearance, appearance as surface. Their surface might be a mask, they might be masqueraders, and they might remain elsewhere (Irigaray 1985a: 76). Their inability to participate wholly in the imaginary of masculinism constitutes both their femininity and their threat. Their mask, their envelope, constitutes both their role and their risk; they are necessary and unwanted guests at the performance. And their mask might indicate 'the abeyance of form, the fissure in form, the reference to another edge' (Irigaray 1991: 56; see also Reichert 1994), a hinge into another spatiality. To mark their presence is to write as if the imaginary space of the master geographer was threatened from within and from without. It is to write as if the mirrors were not solid but permeable, as if the tain could move, as if the glass and silver were melting, as if there was an elsewhere. As if heroes were vampires and as if the women holding hankerchiefs to their faces like shrouds were smiling. It is to write as if the mirrors had bled, bled their violence, bled their ancestry, as if blood could be beautiful, as if an elsewhere was possible.

[The conventionalized need for a conclusion requires a return to an academic voice at this point. Because of course the reverse is also true: the academic voice parodied here requires conclusions.]

In this chapter, I have attempted to displace the opposition between real and imagined space. Despite the recognition by geographers like Harvey, Smith and Katz that the distinction is troubled, they, like many other geographers, continue to insist that it is valid. As Reichert (1992) has also argued, however, theoretical manoeuvres like these themselves assume a particular spatiality. Drawing on the work of Irigaray and Butler, I have described this foundational spatiality as a performance of the male imaginary. It is a spatiality both of boundaries, intervals and solidity, and of flows, fusion and melding; the complex contradictions between these two modalities of space enable but also undermine the conceptual work of distinguishing between different kinds of space. It is this prior spatialized male imaginary which allows geographers to distinguish between real and non-real space in their particular, contradictory manner. I hope thus to have shown that this real space is no more and no less real than the non-real space against which it is defined, and that both entail this imaginary.

And I have tried to do this by masquerading, performing a self, both by articulating different voices and by engaging with Irigaray as my interlocutor in a dialogue. This is not to suggest that I have performed an escape from masculinist discourses and have manipulated an authorial mask from a place entirely outside. Irigaray (1985a: 135) herself insists that such an escape is impossible; she argues that phallocentric discourse must be worked against itself from the inside. Its mirror must be retraversed. This strategy of deconstruction from within means that the risks entailed in her writing, and this performance, are high; masquerade may be assimilable to masculinist discourse as just another example of feminine superficiality. As part of her concern for dialogue and for the intersubjectivity of language, however, Irigaray articulates the importance of the audience to the meaning of a masquerade. So the question of the politics of masquerade might then become, how does the audience interpret the performance: *who are you?* and the answer, *how about you?* This performative context is crucial in thinking about the subversive possibilities of masquerade.[4] It locates the significance of the masquerade not with the masquerader's intentions, but with their relation to their audience, which is to say with the discursive context of the performance of masquerade.

So in another moment of retraversal, I now turn to you as my audience, witnesses to this masquerade, and, as I can't hear your reactions, end this chapter with the recognition that without you, I'm nothing.

ACKNOWLEDGEMENTS

I'm most grateful to the Department of Geography at the University of Syracuse for giving me the opportunity to write and perform this paper. I'd also like to thank Nancy Duncan, Doreen Massey, Matt Sparke, Pam Shurmer-Smith and David Woodhead for their helpful comments on a previous version.

NOTES

1 For an elaboration of these metaphors, see Pratt (1992).
2 It should be noted that elsewhere in their essay, Soja and Hooper (1993) are keen to displace the real/non-real distinction.

3 For some sympathetic discussions of Irigaray's textual masquerades, see Braidotti (1991); Butler (1993: 36–49); Fuss (1989); Gallop (1988: 92–9); Grosz (1989); Whitford (1991).

4 In her recent clarification of her argument about performance, Butler (1993: 12–13) has downplayed its theatrical analogies because, she argues, the interpretation of performance as theatre removes power from the scene by neglecting the anterior regulation of a performance. However, Butler's rejection of the theatrical analogy also erases the question of the audience, or of the social, from her work. Hence she can say little about performative context or about how certain subjects develop 'collective disidentifications' which enable the subversion of the performed norms (Butler 1993: 4).

RE-CORPOREALIZING VISION

Heidi J. Nast and Audrey Kobayashi

INTRODUCTION

This chapter teases out interconnections between spatiality, corporeality and visuality[1] through critically interweaving two very different sorts of works: one, by Jonathan Crary, titled *Techniques of the Observer* (1993), and the other by Carolyn Merchant titled *The Death of Nature* (1983). We have chosen Jonathan Crary because he, perhaps more than any other theorist, explores in detail how visuality has historically been constituted and deployed in *spatial* as well as corporeal ways. In particular, he provides useful graphic descriptions of the main disciplinary practices through which dichotomous notions of subject and object were identified and carried through metaphors of interiority and exteriority. He goes on to argue that this at-once spatial and epistemological divide collapsed in the nineteenth century when subjects became structurally enmeshed into, or part of, machines. For the most part Crary ignores gendered or other corporeal differences. Carolyn Merchant, in contrast, foregrounds gender in an alternative elaboration of the historical emergence of the subject–object split. Specifically, through careful exegesis of historical texts, she argues that the split was mapped onto a Man/Nature (Woman) dichotomy which began to take shape in late Renaissance times and which was instrumental to the rise of 'science', (merchant) capitalism and secular states. Unlike Crary, she does not claim that the dichotomy eventually collapsed, but rather that it became foundational to contemporary social relations. This early work of Merchant remains unique both in her own scholarly repertoire and in the fact that it is the only text on gender and science that demonstrates how an epistemic stance served gendered ends across a number of geo-political fields.[2] While her work might be criticized for indulging in historical over-generalizations and essentialist statements about women, it also points to a very innovative way of understanding the profoundly gendered geo-political effects of particular epistemes.

The chapter begins, then, by outlining Crary's historiographic analysis of vision: how it was understood in scientific terms prior to the early 1800s; how this understanding ruptured in the 1820s and 1830s; and how a new 'physiological' mode of vision largely supplanted the older 'geometrical optics'. Rather than immediately engaging with his historiography, however, we first juxtapose it with Carolyn Merchant's analysis of how Western scientific communities' ways of knowing the world *and the body* changed dramatically in the 1600s; Man and Nature emerged as new, mechanistically interrelated, oppositional categories. The juxtaposition suggests a reframing of Crary's discussion of vision to make issues of gender and other corporealized 'differences' integral rather than incidental or marginal to vision's history. The fourth and fifth parts of the chapter, then, discuss how one might use Merchant's work to re-corporealize Crary's analysis. In particular, her work helps deconstruct his problematic positing of a universal observer, something that requires a singularly positioned and universal (unmarked) body. At the same time, we use Crary's insights to significantly re-work the gendered dualism of Merchant, in the process making both analytical frameworks more malleable to historical constructions of difference.

CRARY'S VISION – PART I:
DECORPOREALIZED TRUTH AND THE MIND'S EYE

In two works titled *Techniques of the Observer* (1993) and 'Modernizing vision' (1988), Jonathan Crary argues that a decisive break in Western ways of visualizing the world occurred in the early 1800s. Before that time, from the late 1500s to the late 1700s, visuality was associated metaphorically and practically with the camera obscura, an important instrument of scientific objectivism and perspectivalism. In the simplest case the camera obscura consisted of a dark enclosed interior in which a person could sit (Figure 5.1). One of the interior walls was punctured by a small opening through which light entered, projecting an inverted image of what was *outside* onto the opposite interior wall. The image was thought to be superior to what was directly observable in the outside world and was relied upon to produce a unifying and ordering image of that world. More importantly, for Crary, the image emanated from the outside and was directly registered on the walls of an *inside*, the image, in a sense, straddling both spatial domains. According to Crary, use of the camera obscura resulted in a phenomenological and epistemological reordering of the world: the universe could now be cast in terms of an interiorized subject and an exteriorized world.

Bordo's (1987) work on Cartesianism and culture suggests that the camera obscura's utility in effecting such a reorganization stemmed from growing and general scientific distrust of the body as both register and mediator of knowledge. It was imperative that one withdraw from the world and the senses, particularly the eye, and cultivate the intellect or 'mind's eye' – an interiorized disembodied arena of 'pure understanding'. For Descartes,

> [t]o achieve this [spatial and bodily] autonomy, the mind must be gradually liberated from the body; it must *become* a 'pure mind' . . . [C]onstant vigilance must be maintained against the distractions of the body. Throughout the *Meditations*, emphasis is placed on training oneself in nonreliance on the body and practice in the art of 'pure understanding.' It is virtually a kind of mechanistic yoga.
>
> (Bordo 1987: 91)

Figure 5.1 Camera obscura, mid-eighteenth century (Crary 1993: 31)

The camera obscura, then, was one means of allowing for a mechanistic giving-over of the senses to an apparatus that would rationally organize and reproduce the world in a way congruent with a disembodied intellect (Crary 1993: 48, 60). Through working in tandem with an interiorized mind's eye, the apparatus was understood to judge and literally trace out transcendent truth. The camera additionally served as a primary terrain for organizing, staging and policing an exteriorized world: the camera could be trained onto this rather than that object; crystals or other objects were employed to filter the light coming into the aperture, and mirrors and different receiving surfaces could be used to re-work the image. The interiorized subject's activities were, however, considered transcendent, carried out in the service of a science that guaranteed 'correspondence between exterior world and interior representation and . . . [that excluded] anything disorderly or unruly' (Crary 1988: 32).[3]

Through numerous published scientific works and experimentations and, secondarily, through its instrumental use in painting and copying, the camera obscura became the dominant model for understanding vision and representation. Accordingly, the observing subject or I, ensconced in the camera, became the metonymic embodiment of an ironically disembodied, transcendent mind's eye. In so far as the interiorized site was divine (it was here that God's ordering of the world was made clear), so too was the ensconced subject (see Bordo 1987).

CRARY'S VISION – PART II:
CORPOREALIZED VISION AND THE EYE

Crary suggests that the camera obscura model of visuality and truth ruptured in the 1820s and 1830s to be largely replaced by a scientific physiological one. The presence of a discerning mind's eye and the explanatory powers of geometrical optics, upon which the camera obscura had depended, were now questioned. The camera obscura, once 'paradigmatic of the dominant [stable, fixed] status of the observer in the seventeenth and eighteenth centuries' was replaced by optical devices such as the phenakistiscope, diorama and stereoscope, signalling a paradigmatic shift in how the observer was located and constituted phenomenologically and epistemologically.

In the case of the phenakistiscope, an observer stood in front of a pie-sized disk or plate having 8–16 slits radiating symmetrically outward from the centre. Attached to, but slightly separated from, the disk via a shaft was a second and parallel disk of matching size. Upon this surface a figure was drawn at sites perfectly opposite the slits, at the same scale but in slightly different positions. As the observer stood in front of the first disk, looking into one of the slits, the two-disk apparatus was spun so that the images were seen in rapid staccatic succession. What would be single stationary figures had the disk been moved very slowly, were now seen to be a single figure in gradual, continuous motion. This and other similar mechanical productions of illusory motion and time emerged out of physiological research on retinal after-images, images that remain in the eye for short period of time after the eye is closed or a light source is removed.

Observational immobility of a subject was similarly required in relation to dioramas, large stage-like settings onto which observers were placed and through which they were carried via an often circular, moving platform. Such a device made observation of numerous perspectives of light and scenery compulsory. Lastly, the binocular stereoscope – similarly developed through physiological research – trained each of the eyes of an (again) immobile observer onto separate and nearly identical side-by-side (or sometimes slightly overlapping) images. The resulting stereographic or three-dimensional image was considered a superior way of representing proximate objects in that it reflected the physiological binarism of optical processes; views from two different optical angles were melded by the brain onto a single virtual plane. More than any other optical device, the stereoscope conflated tangibility with visual experience, conflating the real with the optical (Crary 1993: 124).

For Crary, what is important in all these optical devices is that their workings *required* the observer to become physically integrated into the apparatuses themselves. Accordingly, the observer becomes 'at once a spectator, a subject of empirical research and observation, and an element of machine production' (Crary 1993: 112). It is this triadic, contradictory position of the subject which Crary suggests *collapsed* distinctions between spectacle and surveillance, counter to Foucauldian claims of their separation.

Through the widespread and popular deployment of these and similar apparatuses the viewing subject became divorced from notions (and expectations) of exactitude and correspondence between image and world. Besides the creation of these leisure items (consumed on massive scales in parlours, theatres and street corners), however, the physiological model of vision served more practical ends; most strikingly, it was instrumental to the mechanizing needs of industrial capital. The eye was studied, for example,

in terms of reaction time and thresholds of fatigue and stimulation [which] was not unrelated to increasing demand for knowledge about the adaptation of a human subject to productive tasks in which optimum attention span was indispensable for the rationalization of labor. The economic need for rapid coordination of hand and eye in performing repetitive actions required accurate knowledge of human optical and sensory capacities.

(Crary 1988: 37)

In disrupting previous received notions of a truth situated in a disembodied mind's eye, then, the scientific relocation of vision in a physiological eye discursively and practically severed the observer from correspondent and visually verifiable 'truth' in the material world. The viewing subject, decoupled from a formerly geometrically mediated and stable 'out there', was now considered materially coextensive and mechanistically enmeshed with the world. Crary, drawing on Baudrillard's work on the destabilization of signs and codes beginning in the Renaissance, Benjamin's writings on the *flâneur* and modernity's proliferation of signs and increased consumption, Debord's work on spectacle and Foucault's notion of a 'technology of individuals' in the nineteenth century, suggests that physiological optics recast the observer in terms of a spatially unfixed, and thus more manipulable, subjectivity. The subject, along with a constitutive constellation of viewable objects, was simultaneously everywhere and nowhere. The observer could now be constituted, disciplined, normalized and circulated as an atomistic, interchangeable site of production and consumption. Severed from founding, spatially stable referents, the subject was easily inserted into a hyper-commodified world of signification controlled ultimately by capitalism:

[T]he imperatives of capitalist modernization, while demolishing the field of classical vision, generated techniques for imposing visual attentiveness, rationalizing sensation, and managing perception. They were disciplinary techniques that required a notion of visual experience as instrumental, modifiable, and essentially abstract, and that never allowed a real world to acquire solidity or permanence. Once vision became located in the empirical immediacy of the observer's body, it belonged to time, to flux, to death. The guarantees of authority, identity, and universality supplied by the camera obscura are of another epoch.

(Crary 1993: 24)

THE BODY AND THE MACHINE

The Death of Nature by Carolyn Merchant (1983) in many ways places Crary's concern with how technologies of vision inform epistemological and phenomenological understandings of the world in a larger and more differentiated bodily and geographic context. Rather than concentrating on how discrete models of vision mediated (a universal) subjectivity *per se*, Merchant interrogates how historically differential markings of gendered bodies were enmeshed with, and mediated through, larger social and geo-political relations. Like Crary, she suggests that in Renaissance Europe an epistemological rupture occurred that transfigured social relations and that was mediated by metaphors that were lived as well as spoken. Unlike Crary, however, she suggests that this rupture heralded rather than preceded modernity and that it had gendered effects. Her analysis entertains bodily differences whereas Crary's does not.

Specifically, Merchant analyses how and why medieval organic metaphors, which had spatially ordered the world during the Renaissance, collapsed as tensions arose between two opposing political and scientific discourses. In one camp were the *vitalists* who held that all of the natural world (including humans) contains an innate organic life force, an idea that continued up through the monadic philosophy of Leibniz in the early 1700s. The hierarchical, organic notion of a *body politic* was grounded historically in vitalism: a monarch was literally seen to make up the body head that was also the locus of divine right, while various nobility and commoners made up the social body. The latter was composed of unequal parts, each part having different but ordered rights and functions (see also Haraway 1991a: 7–9).

Sixteenth-century social changes precipitated by mercantilism, however, disrupted the organic model. The growth of a market economy produced greater socio-spatial mobility and was associated with a decentralization of powers that produced a swelling of the metaphorical monarchic head. Both changes were uncontainable by organic body images presaging a shift to *mechanistic* metaphors. The human body, no longer an appropriate metaphor of vitalism, became incorporated into a new terrain of metaphors involving The Machine. Thus, in Descartes' *Treatise of Man* (written around 1632–33) the human body is analysed explicitly as a machine, while he compares the cosmos to an energy-saving machine in his *Principles of Philosophy* (1644). The machine also becomes a metaphor for ideal social organization, a solution to social disorder that ran counter to the organicism of the original meaning of the 'body politic'. Such usage is evident in Hobbes's *Leviathan* (1651; Merchant 1983: 204, 209).

Most importantly, for Merchant, practical and metaphorical deployments of the Machine informed an ontological hiving off of male from female, Man from Nature. Nature was constructed in terms of a passive materiality that was called to order by the masculine, rationalizing gaze and hand of science. The dichotomy was materially strengthened as the bodies, workings and judgements of masculinist science became integral to those of non-monarchic states and capital. New machine technologies employed by merchants and mining enterprises, surveys of feudal lands and new accounting procedures, for example, supported the rise of a bourgeoisie and the demise of monarchies. Production, the state and science were cast as masculine domains from which women were increasingly forbidden (see also Bordo 1987; Schiebinger 1989).[4]

In contrast to Crary's historization of a dichotomous interiority/exteriority divide, then, Merchant focuses on the emergence of a dichotomous and mechanistic Man/Nature relation. And while Crary posits that modernity commenced with the rise of a physiological discourse – which collapsed the interior/exterior divide, Merchant posits that modernity was mediated through a Man/Nature dualism which continues to dominate Western cultural relations today. The next section geographically critiques and re-works both analyses to entertain a more complicated political and spatial field. We begin with Crary.

POSITIONALITY AND VISION

Crary's work is problematic for a number of reasons, most importantly because it implicitly posits a universal observer. Writing about the camera obscura, for example, Crary (1993: 38) asserts that during the seventeenth century it becomes 'the compulsory site from which vision can be conceived or represented'. He does not, however, deal with who and why someone is inside the 'box', how the box is positioned, on whom the aperture is trained, or how those in and outside the box are related; what is implied, therefore, is that all 'eyes' (or mind's eyes) equally identify with and construct the dominant discourse. Accordingly, withdrawal into the camera obscura is described in terms of a metaphysics played out upon a generic, universal body:

> [T]he camera obscura is inseparable from a certain metaphysic of interiority: it is a figure for both the observer who is nominally a free sovereign individual and a privatized subject confined in a quasi-domestic space, cut off from a public exterior world . . .
>
> (1993: 39)

'Domestic' places, however, discursively imply protected, female or womb-like sites, something the camera obscura is obviously not. As the site of a metaphorically transcendent being that polices the world, the inner space is more akin to heaven than home – a control room more than a kitchen or hearth. Clearly, entering a camera obscura to contemplate the world (or contemplating entering such a device) entailed some degree of leisure, scholarship, and institutional and personal privilege denied to all but a few men.[5] Similarly, Crary never addresses the fact that the camera obscura did not seal the (male) observer off from the world; rather, from the recesses of control, this figure was able to call upon discursive and disciplinary (exclusionary) forces not available equally across the social field. Thus, being 'inside' a camera obscura-like space was presumably much different if you were painter or painted, servant or master.[6]

The passage above continues:

> At the same time, another related and equally decisive function of the camera was to sunder the act of seeing from the physical body of the observer, to decorporealize vision . . . [T]he observer's physical and sensory experience is supplanted by the relations between a mechanical apparatus and a pre-given world of objective truth.
>
> (Crary 1993: 39)

But while the body may have been abjected at the level of 'truth', at another level, the 'physical and sensory experience' of the judging observer was not supplanted but, rather, given substantial, god-like powers. As Haraway (1991b: 189) states with reference to the disembodied gaze of science, such pseudo-disembodiment is a 'god-trick' (see also Rose 1993b). Locating the judge is additionally important in that it speaks about *who* and/or *what* is rendered an 'observed': how the box is positioned is not serendipitous, but planned,[7] which as we shall show below, makes a difference as to how the field of exteriority is ontologically constructed.[8]

Lastly, Crary does not address how the camera obscura was able to exert such influence over the body politic nor how it mediated particular kinds of political relations. By not distinguishing epistemological from practical effects of the camera, then, Crary problematically re-creates the transcendent mind–body/inside–outside divide and ignores social difference and located-ness in ways that contradict key historical claims in his analysis.

The contradictions are particularly evident when Crary describes the transition from geometrical to physiological regimes of vision. For Crary, physiological optics allowed observation to be enmeshed in an abstract visual field that required no stable point of observation and no necessary or fixed referential relation to the world-at-large. Corporeality and time were reduced to visual simulacra, allowing desire to be manipulated and mediated visually in ways congruent with the commodifying aims of capitalism.[9] What Crary does not take into account, however, is that the observer is only unlocatable with respect to a particular constellation of objects. A specified ensemble and relation of subjects and objects are, by his accounts, needed to activate (in Foucauldian language, deploy) the physiological mode of vision. The viewer needs to pick up and use a stereoscope, for example, before the physiological mode is enacted or lived.

A corollary to this is that Crary does not question whether or not the necessity or desire for correspondence and coupling between observer and observed varies across socio-political fields: how universal or comparable are, for example, the lives and subjectivities of those enmeshed in a semi-roboticized factory assembly line and those of astronomers, architects, cartographers or surveyors? Is not the former embedded in more bodily and consistent ways with the physiological mode of apprehending the world whereas the others consistently draw upon the geometrical mode of knowing characterized by the camera obscura? And what about those located in rural farm communities or those who refuse to mass consume out of choice or necessity? That Crary does not entertain such questions ironically shows that he has assumed the uninterrogated position of the interiorized observer whose gaze is focused on the 'out there'.

It nonetheless seems clear that while capitalism requires human insertions into mechanical means of production and that it encourages the proliferation and consumption of ever-new commodities and signs registered primarily through images, it is also propelled by a desire to more accurately locate, specify and characterize labour, markets, lands, and materials. The latter desire is, moreover, mediated at a distance through a seemingly transcendent logic of capital accumulation which depends both on policing 'correspondence' of commodified image and object and on maintaining privileged and, typically, masculine, spaces of interiority. It is within these places that 'the logic' (of production and marketing, for example) are tabled, discussed and at some level worked out.[10]

It is here that we re-introduce and re-work Merchant's historical treatment of the Man/Nature dichotomy to suggest that Crary's geometrical and physiological visual regimes are in fact different modalities of the same mechanistic model of objectivism. Obversely, we suggest that by incorporating notions of interiority and exteriority into Merchant's Man/Nature dichotomy, her social location of gender difference significantly changes.

OBJECTIVISM AND BODILY DEPLOYMENT OF THE MIND'S EYE

We have argued that Crary's work presents us with an interiorized subject of the camera obscura who was – as bearer of the mind's eye – representatively transcendent and even divine; he was also male. Drawing upon Merchant's work we suggest that it was exactly this masculinist transcendence which was instrumental in forging mechanistic discourses of, and interrelations between, science, capital and the state. Nonetheless, we would argue that the interiority–exteriority divide is *not* equivalent to that of Man–Nature.

Discursively, for example, the interiorized god-subject is marked ostensibly by an ungendered, universal logic. And (like the Wizard of Oz) the subject is disembodied to the extent that he is hidden from view.[11] Rarely seen patriarchs, for example, are written in as rulers of Bacon's utopian *New Atlantis* (1624). Carried around in gilded carriages, the so-called 'Fathers of Salomon's House' had jewelled fingers, priest-like robes and eyes intriguingly shielded from view by Spanish helmets. One of their missions was to recreate Nature under laboratory conditions according to a mechanistic hierarchy of apprentices, novices and scientists (Merchant 1983: 180–4).[12]

Within the camera obscura set-up, however, these man-god sites of transcendent logic are decidedly distinct from the physical, discursive locations of so-called *Man*. Man, we would argue, is positioned in the field of *exteriority* at some ontological distance from a similarly exteriorized but feminized Nature (Figure 5.2). We would argue further that it was (and is) in this exteriorized field of Man–Nature that physiological research was (and is) located and practised. Taking the cover photograph of Crary's *The Techniques of the Observer* as an example – a close-up drawing of a man's head held firmly by five hands of science, all of which are probing and measuring the curvature of one of the man's eyes – one might say that physiological optics did not supersede, but was epistemically and practically embedded within, the geometrical regime of vision; experimentation, regulation and standardization was for the most part carried out by god-like logicians upon corporeal (not-god) others – 'out there'. In metaphorical terms, then, our argument requires that not only those enmeshed in the dioramas, stereoscopes or phenakistiscopes be considered, but those who designed, produced and/or operated them.

Furthermore, we contest Crary's (1993: 35) claim that physiological research on the eye heralded scientific 'excitement and wonderment at the body', arguing that he conflates excitement with the eye with excitement at the body: mechanistic constructions and treatments of the eye were in fact presaged in much earlier, similar treatments of bodies. Royal College physician

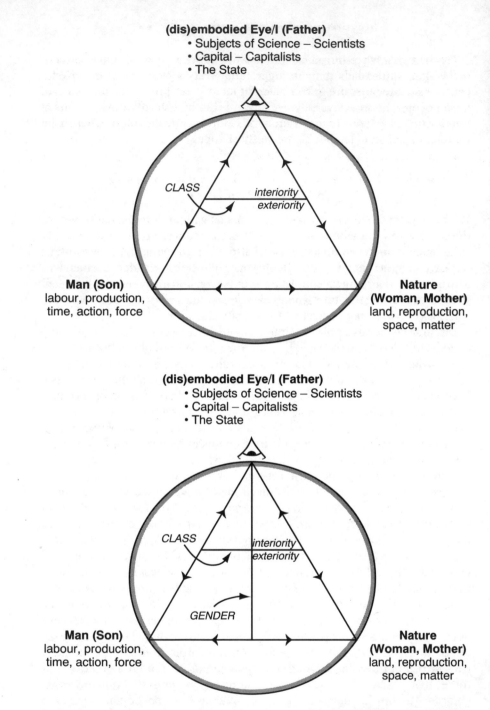

Figure 5.2 Socio-spatial positionings that at once define a particular episteme (objectivism) and geo-political relations. Whereas Merchant focuses on the Man/Nature dichotomy, Crary elaborates upon one defined through an interiorized subject and an exteriorized object. As we discuss in the text, however, their works suggest that there are *two* masculinities, both of which are overdetermined: one which occupies god-like positions of rationalizing authority (Father) and the other which is (primarily through class) positioned at some distance in the field of exteriority, defined geo-politically and

William Harvey, for example, conducted invasive experiments on the heart in the 1630s, consequently comparing it to a pump. He likewise dissected large numbers of deer on the estates of King Charles to study the mechanics of biological reproduction, a logic he then transferred to humans (Merchant 1983: 156–62). Thus, while physiological research of the nineteenth century may have elaborated upon a mechanistic treatment of the body (cf. Crary 1993: 88, 93–4, 148), it did *not* signal an epistemic rupture with previous scientific paradigms or ways of gathering knowledge.

Similarly, mechanistic synchronization of eye movement in the service of capital was preceded by, and logically congruent with, attempts to create mechanistic socio-spatial divisions of labour. Clothier capitalists of seventeenth-century England, for example, divided sorting, carding and spinning tasks of peasant cottagers from those of weaving, dying and dressing of cloth, not only in labour terms, but spatially (Merchant 1983: 177). Analogously, elaborate and mechanistic divisions of labour were drawn and described by Agricola with respect to mining and metallurgy in the sixteenth century (Hoover and Hoover, 1950), while British colonial capitalists in the Caribbean set up the first slave-based factory system of sugar production in the 1600s (Mintz 1985). Moreover, European armies were reorganized explicitly through metaphors of the machine following the success of the Prussian army in the Seven Years War (1756–63; Mitchell 1988; scc notc 4). In this scnsc, the eye was drawn into a process of regulation, experimentation and normalization that had begun much earlier at the scale of the entire body and body politic (see also Braverman 1974) .

Finally, while Crary makes powerful arguments for how the camera obscura mediated interiority, vision and knowledge, Crary treats the camera as paradigmatic of interiority, privacy and knowledge in general. But as Wigley (1992: 347) points out, '[t]he first truly private space was the man's study, a small locked room off his bedroom which no one else ever enters, an intellectual space beyond that of sexuality', such rooms becoming 'a commonplace in the fifteenth century [in Europe]' (1992: 347). Additionally, devices in widespread use such as the astrolabe, quadrant and plane table provided greater mobility to the interiorized mind's eye, extending and inverting the spatial logic of the camera obscura unevenly across mercantilist terrains (see Cosgrove 1985). Here, the discerning mind's eye was expected to measure the world without an image first being registered on some distant surface – at least not in its entirety. In addition, the body stood *outside* the devices rather than being enclosed by them. Nonetheless, these devices, like the camera obscura, were (and are) grounded in a geometrical optics that assumes the presence of a rationalizing, singular observer who judges and polices the truth.[13] One might

---

epistemically (Son). The figure thus illustrates how heteropatriarchy involves an assertion (externalization) of paternal authority over *both* wife (Woman, Mother) and son (Man). Directional arrows represent geo-political forces and tensions constructing and questioning the epistemic order. For the sake of our argument, we have shown only the dominant forces involved in stabilizing these categories, but they by no means make up the *entirety* of socio-spatial relations!

even argue that surveying, as the basis of architectural drawing, urban and regional planning and various kinds of mapping, has since the fifteenth century been of greater importance in shaping our ways of knowing and negotiating the world than the camera obscura (see Cosgrove 1985; Harley 1989).[14]

Rationalizing visual divides placed in the 1400s and 1500s between painter and painted produced a similar kind of distance, if not interiority. These include the orthogonal grid of Brunelleschi (1425) placed between painter and painted. Similarly, Albrecht Durer proposed that a glass plate divide the two and that the painter's head be set in a chin rest to make the painting process more exact by immobilizing the head. Alberti also 'recommended a specially woven veil – called an "intersection" – which when worn over the painter's eyes, "divided" the visual field into square sections' (Bordo 1987: 64). Lastly, Berkeley in *The Theory of Vision Vindicated* (1732) described how perspective could be conceptually captured through imaging a gridded 'diaphanous plain' positioned at right angles to the horizon (Crary 1993: 55).

Thus, if we reconsider Crary's and Merchant's analyses in terms of a more fragmented socio-spatial terrain (especially in terms of gender and class) and in terms of a variety of geometricizing devices that historically promoted ontological distance and interiority, it becomes obvious that the geometrical optics undergirding the camera obscura are still fundamental to contemporary ways of knowing. Founding referents, correspondence between images and objects, and the production of discrete, locatable sites of 'subjects' and 'objects' continue to be fundamental to the 'transcendent' rationalities of capitalism, science and states (see also Lefebvre 1991). Laboratories, penthouse board-rooms and Founding Fathers' (governmental) chambers become contemporary camera obscura-like rooms wherein logics that command and police correspondence between image and world, signifier and signified, are lived out and circulated. 'Targeting' niche markets, capital 'penetration' of new and especially rural areas, inter/transnational capitalist production modes and investment profiles are all policed, judged and managed through techniques underpinned by geometrical optics (Figure 5.2). Facts and figures, graphs and photographs, surveys, quotas and forecasts are means for locating, registering and controlling outside truths in rationalizing fashion.[15] And even if constellations of objects (human and non-human) are captured through commodification and practically and symbolically intermeshed and controlled, total control is continually resisted through consumption. Consumers wrest meaning out of mass-produced objects, objects which Baudrillard – assuming a position of disembodied observer – can only script in terms of an emptiness singularly determined by capitalism. In other words, while he defines mass-produced series of objects as 'undefined simulacra of each other' (Crary 1993: 12), such objects can also be consumed in ways that create new meanings and material referents where none existed before.

What our discussion suggests, then, is that Crary's physiological regime of visuality is in practice deployed and circulated very differently across social fields. Through consumption (of commodities, of scientific rationalities or inventions, and of state representatives) or through certain forms of production (assembly lines), for example, bodies are psychically and materially constituted and mechanistically enmeshed within an ahistorical objective present. The present is in this case mediated by an ever-accelerating circulation of visual (and virtual) images and signs. In contrast, those involved in hegemonic forms of marketing, finance, science or politics are invested in visually, corporeally, spatially (in short, epistemically) distancing – at the same time that they locate and capture – others (cf. Rose 1993b; Haraway 1991b). Such activities to a large extent depend upon a geometrical optics only partially captured by the workings of the camera obscura. The divide is not at all absolute, and the scale and characteristics of the visual regimes vary, but it is important to recognize the socio-spatial unevenness of the regimes and their interconnectedness in contemporary social relations.

In so doing, we open up analytical and practical possibilities for analysing other kinds of corporealities, subjectivities and eyes, especially in fields presently constituted as exterior or 'other'. Most importantly, the exteriorized field of our triangular model could be reworked to include racialized/ethnicized others who have characteristically been cast as both Man and Nature (Beasts or Monsters), depending on economic, political and cultural conditions (Figure 5.3; see also Collins 1991). Slave women in the southern United States, for example, were considered mothers but also beasts (breeders); simultaneously they might occupy masculinized positions as manual labourers in cotton fields or feminized sites in the master's household carrying out domestic functions having the least social status.

CONCLUSION

This chapter has counterposed the works of Crary and Merchant to tease out interconnections between spatiality and visuality, in the process showing how corporeality is multiply constructed, deployed and experienced across a variegated social field. Additionally, it shows that the *epistemic* framework of objectivism is equivalent to, and thus sustained by, normative material positionings through which we are 'named', experience and give meaning to the world. It is in the sense of linking the political and the epistemic, position-ality and the bodily, and the spatial and the visual, that we re-corporealize vision.

We also showed that two taken-for-granted binaries in contemporary Western scholarship (inside–outside and Man–Nature) do *not* in fact map directly onto each other. Rather, through critically interrogating the *bodies*, *visual regimes* and *spatialities* involved in their geo-political construction, the binaries map out a pyramidic socio-spatial terrain, within which are situated two very different sorts of masculinities – Father and Son. Our analysis thus complicates binary analyses of patriarchy. It also suggests an overdetermined and heterosexualized subtext: hegemonic relations within Western socio-spatialities are structured by the Oedipal triad Father–Mother–Son, which is in turn dependent upon a realm of the repressed (Monsters). The masculinity of the third term (Father) is rarely interrogated because it is characteristically *displaced*; it is hidden within the disembodying rubric of transcendent 'rationalities' and logics, an obscurantism involving strenuous cultural work.

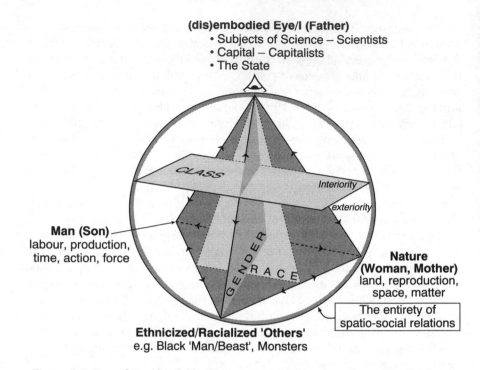

Figure 5.3 Reworking the field of exteriority. Within the white oedipalized order of things (Father–Mother–Son) ethnicized and/or racialized 'Others' occupy an ambiguous position, being drawn into positions of Man, for example, according to economic exigencies. Slave women in the southern United States, for example, were considered both Mothers but also Beasts (breeders) at the same time that they occupied masculinized positions of manual labourers in the fields

Consequently, Man (as labour, maleness, citizen or human object of the scientific gaze) is confused with *Maleness-as-God*, maker and bearer of logic (or worse, Logos itself). Nonetheless, *we* locate and name the third term (as Dorothy and Toto did the invisible Wizard), in the process revealing merely another tier of the masculine, albeit one that is purposefully de-gendered (science is neutral, capital is a logic without a gender, state Fathers represent justice and are for and of 'the people').[16] Distinguishing between modalities of masculinities has important implications for resisting the dominant order of things. It challenges us to interrogate in spatial ways the differences between masculinities of Father and Son and to confront heterosexuality (hetero-patriarchy) rather than the vaguer (safer) notions of patriarchy. In this way we begin to describe the socio-spatiality of Western male–male relations in terms other than simply contradictory, variable or complicated.

Through re-corporealizing vision, then, we have mapped out hegemonic sexualizations of the socio-spatial, a process that simultaneously shows the architectural artifice of the latter's force and which supports those who visualize and act upon a 'beyond'. Such 'going beyond' is not new but an intrinsic part of the dominant order of things; it is that which is continually regulated, de-legitimated, and erased from normative view. Overcoming erasure does not require new practices, then. Rather, it requires acknowledging, working with and supporting differences taking place outside the dominant pyramidal paradigm. Those who have structurally occupied marginalized fields of *exteriority*, for example, have also claimed and re-worked their outside place in terms of an *interiority*, in the process destabilizing elitist and exclusionary private parts. The nineteenth-century American slave woman Harriet Jacobs concealed herself inside a cramped attic room for nine years to escape her master's advances. Her rationalizing gaze through a keyhole was mediated by a fear through which was borne an altogether different sense of interiority. Nonetheless, it was from this different interior that she successfully planned her escape (Jacobs 1988). Obversely, bell hooks's (1990) work shows how marginalized groups and places can inculcate psychical and material interiorities that are oppositional (see also Rose 1993b).[17]

Interiority and exteriority, then, are complexly refracted across social fields and change over time and in direction. Even though metaphors of interiority/exteriority may underpin an inadequate and exclusionary epistemic divide, causing us to seek alternatives (see Grosz 1994), the binary can in another sense be seen as innocuous. It is neither complete (in terms of the hegemonic and heterosexualized (triadic) social terrain which it only helps to define) or absolute (see also Haraway 1991b: 194; Rose 1993b). Its instabilities are, moreover, especially prominent today. Global structural changes have opened up numerous possibilities for recognizing, legitimating and supporting alternative socio-spatial relations (epistemes) outside the dominant heterosexualized order of things. We are today confronted with numerous possibilities for seeking 'not the knowledges ruled by phallogocentrism (nostalgia for the presence of the one true Word) and disembodied vision, but those ruled by partial sight and limited voice' (Haraway 1991b: 196). Our project here is thus one of disclosure – of locating and speaking about the socio-spatial pervasiveness and artifice of heterosexuality, one underpinned by two very different sorts of masculinities.

ACKNOWLEDGEMENTS

We would like to thank the anonymous readers, Mike Dorn, Angela Martin, Alexis Papadopoulos, Susan Roberts and Greg Waller, for their close, critical readings of the text. Thanks also to Nancy Duncan for organizing the Space, Place and Gender seminar series out of which this collection grew. Lastly many thanks to Dick Gilbreath and Gyula Pauer for their patience, skills and creative inputs in the making of the figures.

NOTES

1 The term 'visuality' implies that there are different ways in which the visual field is, and has been, culturally, epistemically, materially and historically apprehended and organized – all of the latter being interconnected. Thus, persons who do not operate within visual regimes structured predominantly by perspectivalism, for

example, *see* the world differently. Visuality also implies that there are differences in the techniques and technologies used to support particular ways of seeing and thereby also in the power relations which undergird, and are dispersed through, these differences.

2 By 'geo-political field' we mean all sites within, across and through which power is materially deployed. Thus, the body is one geographical site, but so are various European sites. Moreover, those material and discursive apparatuses which are used to code and mediate socio-political relations but which are often less easy to 'place' are included here, for example, the 'state', 'science' and 'capital' – all of the latter ultimately being locatable.

3 Crary (1993: 34) is careful to caution that the ontological distinctions that emerged through the metaphor and use of the camera obscura should not be conflated with the technique of linear perspective itself, in which the world is framed according to particular visual rules:

> one must be wary of conflating the meanings and effects of the camera obscura with techniques of linear perspective. Obviously the two are related, but it must be stressed that the camera obscura defines the position of an interiorized observer to an exterior world, not just to a two-dimensional representation, as is the case with perspective . . . [I]t is far more than the relation of an observer to a certain procedure of picture making . . . [T]he phenomenological differences between the experience of a perspectival construction and the projection of the camera obscura are not even comparable. What is crucial about the camera obscura is its relation of the observer to the undemarcated, undifferentiated expanse of the world outside, and how its apparatus makes an orderly cut or delimitation of that field allowing it to be viewed, without sacrificing the vitality of its being.

This passage nevertheless becomes problematic when he asserts that the camera obscura was singularly effective in making objectivism the dominant epistemological mode of enquiry.

4 See Mitchell (1988) for a fascinating account of how machinic metaphors and practices completely re-worked Egyptian society which had formerly been structured through organic metaphors of the body. Beginning with the Ottomans, who adopted Prussian military techniques after the French invasion in 1798, for example, the (all-male) military became:

> a piece of machinery that could be 'tuned with the precision of a watch.' It could be made to perform what the French officers in Egypt now called 'manoeuvres,' to rotate, discharge weapons, contract, or expand on command . . . In such a machine, every individual occupied a position, a space, created (as with the cog of a wheel) by the identity of interval between each one . . . Order was a framework of lines and spaces, created out of men, in which men could be distributed, manoeuvred, and confined.
>
> (Mitchell 1988: 38)

Adoption of machinic ways in future colonial settings complicates both Merchant's and Crary's analysis in that it shows the ways in which machinic ways were dispersed for specifically military purposes. Moreover, it shows how persons were embedded within machine-like contexts (what Crary, in the context of vision calls the 'physiological mode') much earlier than the 1820s and for different reasons. Mitchell (1988: 37, 38) writes, for example, that:

> [following] the dramatic Prussian victories in the Seven Years War (1756–63) . . . [t]he Prussians had introduced revolutionary [military] techniques of precise timing, rapid signalling, and rigorous conformity to discipline, out of which an army could be manufactured as what the Prussian military instructions called an 'artificial machine' . . . Such a 'machine' could fire with a rapidity three times that of other armies, making it three

times as destructive, and could be expanded, wheeled, and withdrawn with mechanical ease . . . The Prussian military regulations were adopted in the decades after the Seven Years War by all the major armies of Europe, and improved upon by the French in new regulations of 1791.

The officers of the new order . . . could 'dispose of a large body of men in a circular form, and then cause them to march round in such a manner, that as the circle turns the soldiers incessantly discharge their muskets on the enemy and give no respite to the combat, and having prepared their guns for a fresh discharge before they return to the same place, they fire the moment they arrive in the face of the enemy. The result of this circular formation is, that the fire and slaughter do not cease for an instant.'

5 Another instance of gender blindness occurs when Crary describes in universal terms how, for Leibniz, to be *in camera* in the 1600s meant to be in the 'inner space' of a judge or other person of title. He ignores the strikingly gendered class aspects of this scenario (judges of the 1600s were male and of high social rank) and merely posits a universal (male) figure. His discussion of how the camera obscura infused popular notions of a room, therefore, does not recognize that such rooms discursively and practically helped construct an elite male figure.

6 To some extent class and gender aspects are touched upon when he writes that, '[f]or those who understood its optical underpinnings it [the camera obscura] offered the spectacle of representation operating transparently, and for those ignorant of its principles it afforded the pleasures of illusion' (1993: 33). But who was ignorant and who was knowledgeable is not explicitly addressed, nor are the political and epistemological implications of this distinction explored.

7 Crary (1993: 34) to some extent acknowledges this point, but in an abstract way:

What is crucial about the camera obscura is its relation of the observer to the undemarcated, undifferentiated expanse of the world outside, and how its apparatus makes an orderly cut or delimitation of that field allowing it to be viewed, without sacrificing the vitality of its being.

What Crary does not question is the 'cut' itself: on what or whom is the linear gaze trained and what difference would this make to constructions of the visual field? He also does not question whether there are differential qualities to exteriority and interiority that help maintain or transform the objectivist order of things.

It is improbable, for example, that Vermeer's paintings of *The [male] Astronomer* (1668) and *The [male] Geographer* (1668–9) solemnly poring over charts and maps next to a single window in a camera obscura-like room (paintings reproduced and discussed in Crary's book) invoked the same sense of inside and outside in women and in men. There were ideologies of femininity and masculinity as well as of class that informed the looking, thereby enabling different social groups to relate and mediate their social positions (of insiderness/outsiderness) differently yet simultaneously. Moreover, men of lower social rank would have lived out their lives feeling a greater sense of exclusion from the places of the juridical eye, perhaps living more fully within the domain of exteriority. Lower-class women perhaps also experienced a greater sense of exteriority/exclusion, but in ways different from those of men, that is, according to constructions of gender.

8 Haraway (1991b: 193–4), writing with reference to Western science states that, '[p]ositioning is . . . the key practice grounding knowledge organized around the imagery of vision' and that:

Technologies are skilled practices. How to See? Where to see from? What limits to vision? What to see for? Whom to see with? Who gets to have more than one point of view? Who gets blinkered? Who wears blinkers? Who interprets the visual field? What other sensory powers do we wish to cultivate besides vision? Moral and political discourse should be the paradigm of rational discourse in the imagery and technologies of vision.

Similarly, Haraway (1991b: 193) writes that '[v]ision is *always* a question of the power to see' and that:

> [the] eyes have been used to signify a perverse capacity – honed to perfection in the history of science tied to militarism, capitalism, colonialism, and male supremacy – to distance the knowing subject from everybody and everything in the interests of unfettered power.
>
> (1988: 198)

9 In capitalism, series of things are produced which appear identical such that:

> [t]he relation between them [identical objects] is no longer that of an original to its counterfeit. The relation is neither analogy nor reflection, but equivalence and indifference. In a series, objects become undefined simulacra of each other.
>
> (Baudrillard quoted in Crary 1993: 12)

10 We purposefully use the word 'table' to draw attention to its epistemological links with the camera obscura and especially its relation to the Cartesian table. Quoting Foucault, Crary (1993: 56) writes that, 'The center of knowledge in the seventeenth and eighteenth centuries is the *table*.' In the case of one of Chardin's eighteenth-century paintings, the table figures prominently and has 'less to do with sheer optical appearances than with knowledge of isomorphisms and positions on a unified terrain' (Crary 1993: 63). We argue, then, that the Cartesian table and geometrical optics still predominate today, whether they are used to register and/or prove particular logical points in mathematics, cartography (including Geographical Information Systems), corporate boardrooms, banks and stock exchanges or governmental chambers.

11 Again, Haraway (1991b: 193) writes similarly, but in the context of a critique of science: 'Only those occupying the positions of the dominators are self-identical, unmarked, disembodied, unmediated, transcendent, born again.'

12 Analogously, the forerunner to the Royal Society was the Invisible College, a name connoting bodily transcendence and truth.

13 For a particularly relevant and contemporary example of continued scientific policing, see Harley's (1989: 4–5) discussion of various forms of cartographic vigilantism. A French edited collection, *Cartographie dans les médias*, for example, is said to 'attempt to exorcise from "the realm of cartography any graphic representation that is not a simple planimetric image and to then classify all other maps as 'decorative graphics masquerading as maps' where the 'bending of cartographic rules' has taken place . . . "' He also states that 'in Britain . . . there was set up a "Media Map Watch" in 1985', an 'example of cartographic vigilantism [in which] the "ethic of accuracy" is being defended with some ideological fervor . . . The best maps are [judged to be] those with an "authoritative image of self-evident factuality"' (all quotes from page 5).

14 Indeed, as Harley (1989: 1) points out with respect to cartography, 'Applying conceptions of literary history to the history of cartography, it would appear that we are still working largely in either a "premodern" or "modern" rather than in a "postmodern" climate of thought . . . [W]e are still, willingly or unwillingly, the prisoners of our own past.' For Harley, cartographic pre-modernity is carried through staunchly defended disciplinary rules which assume that 'objects in the world to be mapped are real and objective, and that they enjoy an existence independent of the cartographer; that their reality can be expressed in mathematical terms; that systematic observation and measurement offer the only route to cartographic truth . . . [M]imetic bondage has led to a tendency not only to look down on the maps of the past (with a dismissive scientific chauvinism) but also to regard the maps of other non-Western or early cultures . . . as inferior to European maps.'

15 A particularly graphic contemporary example of the omnipotent and rationalizing position of capital is found in Natter and Jones's (1993: 141) article which in part

describes the post-recession, industrial 'restructuring of Flint's [Michigan] space into ever more distinct public and private spheres' following General Motors' devastating plant closures which resulted in the loss of 30,000 jobs. One indication of the resultant dichotomization was 'that the territories of the privileged, e.g., corporation headquarters and private clubs, are policed to ensure that opposition is held at bay'. We would argue that this policed distancing not only allows opposition to be managed, it disallows *viewing* of corporate capitalists' bodies, in the process promoting the myth of corporate omnipotence and reproducing a contemporary version of a mind's eye guarding a rationalizing logic bound to profit.

16 It would be interesting to see how our analysis might re-work the spatial dualisms feminist geographers have typically posited and interrogated. Drawing on a number of feminist theorists, for example, Rose (1993: 74) lists some of the most common binaries typically held up as constitutive of gendered differences.

17 Rose (1993b: 151–2) might describe such contradictory constructions of interiority in terms of 'paradoxical space' where:

> The Other is not outside the discursive territory of the Same. Eve Kosofsky Sedgwick has explored this paradox in the case of the homosexual and the trope of the closet; she suggests that the image of the closet represents homosexuality as an open secret around which a certain knowledgeable ignorance can centre. Diana Fuss has also pointed out the complexity of this doubled position for gay men and lesbian women. Their simultaneous inside-ness and outside-ness produces many unpredictable paradoxes: for example, 'to be out, in common gay parlance . . . is really to be in – inside the realm of the visible, the speakable, the culturally intelligible . . . [but] to come out can also work not to situate one on the inside but to jettison one from it.' These paradoxes of 'inside' and 'outside' can painfully disempower those caught within them but, as Fuss also argues, 'one can, by using these contested words, use them up, exhaust them, transform them into the historical concepts they are and always have been' . . . [In like fashion] Patricia Hill Collins . . . describes the simultaneous occupation of a position both inside and outside the centre as the 'outside-within stance', and she suggests that it is a position articulated very often by black women because of their role as domestic workers in white homes. There they were on intimate terms with the children of the family . . . but were also made to know that they did not belong . . . Frye suggests a similar subject position for white lesbians and gays, but one which is enabled for different reasons; by acting straight they can be inside but also watch as outsiders.

It would also be interesting to explore why the masculinity of the Father is interiorized and the femininity of Mother is exteriorized (in spatial terms) at the same time that the Father is deemed phallic and the Mother castrated, reversing associations of exteriority and interiority *at the level of the body*.

PART II

(RE)NEGOTIATIONS

GENDERING NATIONHOOD

A feminist engagement with national identity

Joanne P. Sharp

INTRODUCTION

Despite intellectual narratives that describe the increasing internationalization or globalization of life, *realpolitik* would seem to suggest that nation-states continue to be significant actors in the constitution of international society. National self-determination has been a prevalent source of legitimation in many political struggles and national statehood is a requirement for representation in many global bodies. National identity is a central aspect of contemporary subjectivity and yet, certainly within the discipline of geography, its articulation with gender has been largely ignored (but see Johnson 1995; Marston 1990; Nash 1994). This chapter represents an investigation into some of the genderings of nation in both academic theories of national identity and in the operation of nationalisms in the materiality of everyday life.

The silencing of gendered identities of nationalism is due in part to the taken-for-granted nature of national identity in the contemporary world system. Although the gendering of nationalism has been interrogated from some post-colonial locations (for example Chatterjee 1993; Spivak and Guha 1988) nationalism and gender in the 'First' and 'Second' Worlds has received much less coverage (but again see, for example, Johnson 1995; Marston 1990; Nash 1994). I want to take examples of nationalism and gender in the formerly communist countries of Eastern Europe. Here nationalism is foregrounded as the new communities seek to unite territory around old national and ethnic identities that have been suppressed for half a century, and the tensions between national and gendered identities are very clear indeed.

I will conclude with a short discussion of what my analysis of the relationship between national and gendered communities might say to a radical democratic feminist politics.

THEORIES OF NATIONALISM

'At the origin of every nation', states Geoffrey Bennington (1990: 121) 'we find a story of the nation's origin.' This seemingly paradoxical definition of the nation is central to my argument about the convergence of national and gendered identities. Bennington's statement suggests that trying to find a national essence at the formative point of national history is a hopeless quest. The nation is created not through an originary moment or culturally distinct essence but through the repetition of symbols that come to represent the nation's origin and its uniqueness. National culture and character are ritualistic so that every repetition of its symbols serves to reinforce national identity. In his classic text *What is a Nation?*, Ernest Renan (1990: 19) claims that a 'nation's existence is . . . a daily plebiscite'. Each drawing of maps of nation-state territory, each playing of the national anthem or laying of wreaths at war memorials, every spectatorship of national sports events and so on represents this daily affirmation of national identification. Traditions of ceremony, monument and national celebration have instilled national identity into the calendar and the landscape. National identity therefore becomes naturalized, its creation is hidden so that it becomes an unquestioned facet of everyday life.

Bennington's definition of national identity shares the same post-structural genealogy as Judith Butler's description of the social construction of gendered identity. She argues that gender is constituted not by 'a founding act but rather a regulated pattern or repetition' (Butler 1990: 145). Like national identity, gendered identity takes on its apparently 'natural' presence through the repeated performance of gender norms. In the performance of identity in everyday life, the two identifications converge. The symbols of nationalism are not gender neutral but in enforcing a national norm, they implicitly or explicitly construct a set of gendered norms.

Typically however, intellectual analyses remain silent on the gendered construction of nationalism. Take for example Benedict Anderson's now classic analysis of nationalism, *Imagined Communities* (1983/1991). The nation is imagined, he claims, because it is not possible for all members of any nation to know even a small fraction of the other citizens of the country; and yet nations are communities, very real bonds are perceived as linking distant people in the same territory. Anderson points to the importance of mass-produced books and newspapers in vernacular languages – print capitalism in his terms – in the creation of the imagined community of nation. Books and especially newspapers help to forge this national identity. They illustrate the coexistence of activities, concerns and people within the nation-state territory. Readers become aware of a population which shares their identity. The juxtaposition of stories in the media from various locations within the nation produces a notion of simultaneity – an empty calendrical time – which encompasses the entire citizenship in a rhetorical horizontal bond (Anderson 1983/1991).

Anderson's description of the nation as the product of the imaginations of the citizens of a territory has proved to be a brilliant insight into the origin and spread of modern nationalism. He has highlighted the creativity of nation-building – nations are not entirely invented but constructed out of already existent elements of culture, society and mythology – aspects of the past and aspirations for the future are woven into the fabric of daily life in a territory. However, I would suggest that Anderson's account is unidirectional. In his focus upon the construction of the nation, he silences the reverse process, the construction of the national citizen. For not only is the nation imagined to loom out of an immemorial past and glide towards a limitless future (Anderson 1991: 11–12), but at the same time, it is imagined to be peopled by citizens appropriately cultured for this mission. Although he does not mention this explicitly, Anderson's thesis of imagined community assumes an imagined citizen, and this citizen is gendered.

The national citizen is invoked most vividly in the opening sentences of the second chapter of *Imagined Communities*. Anderson (1991: 9) claims the following: 'No more arresting emblems of the modern culture of nationalism exist than cenotaphs and tombs of Unknown Soldiers.' it is the very anonymity of the soldier, he continues, which insures and represents the mythos of the nation: it could be any member in the fraternity of the imagined community. But surely the Unknown Soldier is not entirely anonymous. We can all be fairly sure that the soldier is not called Sarah or Lucy or Jane . . .

The imagined bonding between individuals and the nation in narratives of national identification is differentiated by gender. Men are incorporated into the nation metonymically. As the Unknown Soldier could potentially be any man who has laid down his life for his nation, the nation is embodied within each man and each man comes to embody the nation. This is the horizontal fraternity to which Anderson refers. Women are scripted into the national imaginary in a different manner. Women are not equal to the nation but symbolic of it. Many nations are figuratively female – Britannia, Marianne and Mother Russia come immediately to mind. In the national imaginary, women are mothers of the nation or vulnerable citizens to be protected. Anne McClintock has observed that in this symbolic role, women 'are typically construed as the symbolic bearers of the nation but are denied any direct relation to national agency' (McClintock 1993: 62). It is the metonymic bond of male citizens who must act to save or promote the female nation. Many commentators have mentioned, for example, women's conspicuous presence at demonstrations throughout Eastern Europe in 1989, and their equally conspicuous absence from negotiating tables and governments ever since (see Einhorn 1991; Kiss 1991).

A number of commentators have recognized the importance of risking one's life for one's country in the construction of national citizenship (see Tickner 1992; Yuval-Davis 1991). This role of good citizenship is not one generally offered to women. Although, increasingly there are women in active military duty, recent media coverage has not represented this trend as progressive. Matthew Sparke describes reactions to American women in action during the Gulf War:

> . . . the way the media focused on women who went away leaving children behind brought to light the most hallowed and . . . significant of American concerns, the demise of the nuclear family. The participation of women also helped make more manifest the masculinity of the war machine in general, with grave faces of concern at women going missing in action, and with men . . . bemoaning the 'feminization of the American military'.
>
> (Sparke 1994: 1072)

Much concern has been voiced over the dangers for women in front-line positions. This was symbolized by reactions to the sexual assault of American army officer Rhonda Cornum during capture by Iraqis. Despite her own professional downplaying of the experience, and growing revelations of American non-combatant service-women assaulted by fellow soldiers at the 'Tailhook' Convention and less publicized occasions,[1] the danger of *foreign* violation of female soldiers is one of the reasons why women ground troops have not been allowed to fight at the front line (Enloe 1993: 223). Prevention of foreign penetration of the motherland – and women's bodies as symbols of it – is at the very heart of national-state security. The female is a prominent *symbol* of nationalism and honour. But this is a symbol to be protected by masculine agency.

THE BODY OF THE NATION

Nowhere is the nation more directly embodied as female than in debates over abortion. In many places, including the re-emerging nation-states of Eastern Europe and the post-Soviet Union, women's bodies and the symbolic body of the nation become significantly enmeshed both discursively and materially in hegemonic nationalist discourse. The safeguarding of life of/in women is consistently written in terms of the security of the nation.

Nanette Funk (1993: 194) claims that 'During the Cold War, competition between East and West Germany had provided an incentive for East Germany to liberalize abortion laws . . . ' In a sense, such liberalization symbolized a modern nation-state unfettered by the demands of religious moralities. Now, as part of the backlash against communist regimes, previous policies concerning morality, traditional family values or religion are being overturned. These emotionally charged issues provided lines of resistance during communist regimes and, in nations with a strong religious (especially Catholic) tradition, are currently proving to be important elements of post-communist national identification (Watson 1993: 472). Communism is perceived to have eroded traditional values and now these values are being recuperated, often in rather uncritical form.

In a number of Eastern European countries at the close of socialist or communist rule, ethnic tensions have made women's reproductive capabilities an issue of national interest, even national survival. In Croatia, for example, the nationalist party outlawed abortion in 1992. Slavenka Drakulić (1993: 125) has explained what she understands to be the logic of the Croatians' decision:

> One cannot expect that such a nationalist party, worried that the Croatian nation is soon going to disappear because Croatian women aren't giving birth to more than 1.8 children, is going to promote anything progressive for women.

In Serbia on the other hand, there were covert ethnic policies in place which operated through limitations to childcare and social security provision for women with more than three children – a family structure characteristic of minority Albanians (Milić 1993: 113). These policies therefore have differential national effects. Here, nationalism is aimed more consciously and overtly towards women's bodies. In the war in Bosnia-Herzegovina, the rape of Bosnian minority women by Serb soldiers – in many cases action ordered of them – solidifies all too clearly the links between individual female bodies and the nation in both nationalist rhetoric and *realpolitik*.

NATION, FAMILY AND FEMINISM

The revival of nationalism across Eastern Europe has made all too clear the tensions between the requirements of national unity and feminist demands. Interpreters of post-communism in Eastern Europe note that, despite cultural and historical differences between individual states and nations, feminism is an identity that is consistently rejected throughout the region. The presence of feminist movements is significant in my argument for the construction of gendered subjects in that feminism represents a self-conscious politicization of gendered identity.[2] I think that it is first important to understand the construction of gender relations under communism in order to explain why it is that feminism does not have the prominence within contemporary nationalist debates there that Western feminist analysts might expect.

In demarcating a distinctive communist and post-communist voice however, I am silencing cultural differences *among* Eastern European states. There is a danger, when talking of characteristics common to post-communist countries, of replicating the East–West geo-political opposition of the Cold War period. There are significant cultural differences among the countries of Eastern Europe and the former USSR, including pre-communist political structure, economic development, degree of liberalization permitted after Stalin's death, linkages between intellectual dissent and working-class movements, influence of non-European cultures and so on. In addition, there are similarities between gender relations in the (post-)communist countries and in the West. However, I believe that there are sufficiently significant common trends in these countries emerging from several decades of communist governance to justify a joint discussion of them. This shared history offers an experience which differs from the scripting of gendered subjectivity in other places.

Feminism has a different social location in European post-communist cultures than in Western Europe. During the communist period, the primary division in society was not public/private *per se* but a division 'between public (mendacious, ideological) and private (dignified, truthful) discourses' (Einhorn 1993: 3). The private sphere of the family provided the space of resistance in opposition to the public space of the state. Václav Havel and George Konrád cynically called this space 'anti-politics' to symbolize rejection of state political action. In essence then the family provided a kind of civil society (Goven 1993). Furthermore, the regaining of a 'traditionally prescribed gender identity is an important aspect of nostalgia for "normality"' which people anticipate current change will provide (Watson 1993: 473).

The domestic does not have the same connotations in Eastern Europe as it does in the West. The commonplace binary which Doreen Massey (this volume) notes as existing between home and work does not apply here: the dualisms immanent/transcendent, emotional/rational, private/public, female/ male do not correspond to distinctions between domestic and work spheres in Eastern Europe. In the domestic sphere, political discussions were conducted around the kitchen table, the home represented the only place for open discussion like this; it was here rather than in any public sphere that the possibility for transcendence occurred. This spatiality is bound to have influenced the reproduction of gendered identities in a different way from those described by Massey.

This alternative division of social spheres has had two significant effects on the development of gendered consciousness in Eastern European countries. First, women's ability to leave the domestic sphere to find work is not the same symbol of liberation as it is for middle-class liberal feminists[3] in the West. Under communism, full employment did not represent a fulfilment of the right to work; work outside the home was mandatory. Women's work was concentrated in blue-collar sectors of the economy (Dölling 1991: 9). Furthermore, in this mode of socialist equality, there was no reduction of women's domestic labour when they left the home to work; they had to shoulder this 'double burden'. Thus, movement outside the home was seen by women ambivalently. Hauser *et al.* (1993: 261) state that '[m]any Polish women cannot decide if their employment under communism marked liberation or oppression'.

Second, because the family occupied the social space posed in opposition to the state, women were able to effect few changes to familial or gender roles which would be accepted as progressive by mainstream society. The family symbolized free space in contrast to the state and this tended to deflect attention away from power dynamics operating within the family. Because of the dualism family/state, women who criticized the family were frequently presented as in collusion with the state and women's internalization of the dualism has made them reluctant to question this division. There has in fact been a devaluation of the language of women's emancipation because of its association with the rhetoric of communist regimes (Anastasya Posadskaya in Molyneux 1991: 135). In effect this has discouraged women from pressing for state intervention in the 'private sphere' or family matters as is the case with many women's movements in the West (Einhorn 1993: 6).

This tendency has continued into the present: the notion of female emancipation is strongly associated with old communist regimes (Jankowska 1991: 174) and, as a result, less attractive to nationalist movements. Feminist issues are either considered leftist (Funk and Mueller 1993) and as such become discredited by association with the old regimes, or perceived as representing an unconditional acceptance of Western values (Fábián 1993: 74). Further, despite many apparently favourable conditions for the emergence of a feminist consciousness across the former communist states, there has not been this expression of identity. Kiss has suggested that this is because emancipation was granted by the state – equality (or at least the rhetoric of equality) was imposed from above – rather than women engaging in social change to achieve their own emancipation (Kiss 1991: 51).

Women are often blamed for social problems. In his book *Perestroika,*

Mikhail Gorbachev (1987: 117) claimed that 'women no longer have time to perform their everyday daily duties at home . . . many of our problems . . . are partially caused by the weakening of family ties and slack attitude to family responsibilities'. Gorbachev, evidently addressing a presumed male citizenry despite the universalistic aspirations of his book, asked, 'what we should do to make it possible for women to return to their purely womanly mission' (1987: 117; emphasis mine).

This attitude legitimates women's return to the domestic sphere and helps to alleviate high unemployment caused by the closure of state controlled industry by nationalist governments. Furthermore, previously state-sponsored provision of childcare and maternity leave are among the first services to be discontinued in the transition to a capitalist economy. As I have already mentioned, this attitude is not held only by men. Feminism is weak in Eastern European countries as it is widely regarded as anti-man or de-feminizing.[4] These feelings are enforced in attempts to unite the community in the face of hardship. Bulgarian feminist Maria Todova (1993) tells of claims which state that since Bulgaria is at such a low socioeconomic level, society cannot 'afford' to become divided along sex or gender lines, as if somehow it were feminism not sexism which split the community. What this represents is a movement of gender issues from the public sphere of contested politics into a private sphcrc of cultural mattcrs in thc manncr that Partha Chattcrjcc (1993: 133) has described during India's struggle for national independence. During the nationalist revolt the 'women's question' disappeared as a political issue, he observes, because this inner domain of sovereignty was removed from the public sphere of political contest with the colonial state. It was argued that the 'women's question' would return to the public sphere of politics once independence had been attained.

Cynthia Enloe, however, has questioned the tactics described by Todova and Chatterjee. There is of course a rationality to the suppression of difference in the name of unity when facing a common foe or struggling toward a common goal. But is it possible for a change in leadership, however revolu-tionary, to facilitate emancipation for all if gender problems are not addressed before and during the revolution (Enloe 1989: 57)? Gender cannot be teased out of other relations of power which constitute individual subjectivity but must be seen to exist contingently in all situations – no one can be without gender and in most social locations this is a powerful aspect of subjectivity. If this is the case then surely the postponement of a consideration of gender issues in the name of the construction of the nation-state will irrevocably alter the direction the emerging nation-state will take. The very nature of national identity will be different depending upon whether or not it deals with gender issues at the outset. National identity, therefore, is not something which can be retrieved intact from the past and slotted back into the heart of contempo-rary culture unchanged, it is constituted in particular times and places through relations of power already existent in society.

Certainly this is a problem currently being faced by nationalisms across Eastern Europe. Maria Todova (1993: 30) describes what she sees as an irony in attempts at moving towards democracy in what was Yugoslavia:

The illusion is that once democracy is achieved women, as part of the body politic, will automatically benefit. This framework recalls classical socialist theory: once socialism is installed, it is said, women will be automatically emancipated.

Antić (1991: 150) has also noted the privileging of the national interest over the interests of 'the so called "social minorities"'. In one sense, for women (as for other marginalized groups), structurally nothing has changed in the move from socialist-communist societies to national ones. Once again one subject dominates social identification – now the national subject is pre-eminent whereas in the previous regime all subjectivities were secondary to the worker (Antić 1991).

The question which immediately comes to mind is why it is that 'one politics [is] to be spoken in terms of an-other politics' (Radhakrishnan 1992: 81). Why is it that the politics of gender relations should be forced into the mould of national identity? Expecting that the achievement of democracy or socialism or national self-determination will automatically produce equality is to rely upon too simplistic a notion of social change. It assumes a binary of good system–bad system and, perhaps even more importantly, assumes that democracy, socialism or national self-determination are static social conditions rather than processes of social transformation. No society is *entirely* democratic, socialist or self-determined; the regulation of social life is over-determined so that a state is constituted by an combination of ideal types.

It is clear that changes in Eastern Europe have been made to the social and economic organization of countries but gender relations have remained relatively unchanged. Presumably one explanation of this could be that these socioeconomic changes are not sufficiently widespread for effects on gender relations to be noticeable at present. Feminist issues will come later. But why should this be so? Will attitudes to women's roles in social life be transformed later so allowing the inclusion of women into the national citizenship proper?

GENDER AND THE NATIONAL COMMUNITY

Iris Young (1990) has suggested that, from an orthodox point of view, communities are based upon an argument which posits the necessity for the privileging of unity over difference. The fear is that the alternative, the recognition of differences within the community would destroy the agency of community identity. But it is not necessary to chose either extreme of unitary identity or immeasurable difference. All subjects are constituted through a plurality of facets, but, in Chantal Mouffe's (1992: 372) words,

> this plurality does not involve the *coexistence*, one by one, of a plurality of subject positions but rather the constant subversion and overdetermination of one by the others, which make possible the generation of 'totalizing effects' within a field characterized by open and indeterminate frontiers.

Although there are no pre-existent or inherent linkages between subject positions, this does not mean that each exists in individual isolation.

Contingent relations are constantly forming, creating bonds and common-
alties between these positions.

Following Mouffe's argument, I believe that transformation to democracy
cannot be successful without the creation of a radical democratic process from
the outset, a process which acknowledges the diversity that must be addressed
in creating equality: sameness in treatment is not equivalent to the creation of
equality. There is a danger that the general good may be valued above that
of any discernible group. Following this hierarchical logic, the general good of
women would come above any particular group of women so that some (sex
workers, for example) will be excluded from the construction of demands. I do
not argue against the social necessity of priorities in social reform, it would be
utopian to believe that everything can be changed at once. But I do believe
that the making of priorities should be open and negotiated rather than
concealed under apparently universal claims of a 'true' national interest and
identity.

This discussion should not be taken to imply that the nation-state embodies
some kind of plot which is attempting to control women in the name of male
supremacy. I have been describing the national construction of women but
this is only part of the constitution of gender relations. The role of men in the
reproduction of national cultures is similarly constructed. Obviously not all
men want to have to prove themselves part of their national community
by aggressively defending national borders, just as there are women who do
support this project. The modern nation-state should be viewed instead, I
believe, as a discursive practice in the manner that Foucault proposed. In such
a practice, power is not entirely concentrated in the hands of a ruling elite but
is diffuse. All subjects, ruler and ruled alike, are constrained by their location
in the discursive networks underwriting society. In Foucault's (1980: 188)
own terms,

> For the State to function in the way that it does, there must be,
> between male and female . . . quite specific relations of domination
> which have their own configurations and relative autonomy . . . Power
> is constructed and functions on the basis of particular powers, myriad
> issues, myriad effects of power.

Nevertheless, in the present construction of gendered relations within the
nation-state, we can see a privileging of masculinity. Quite simply, the position
of men in national rhetoric grants them significantly more agency.

But just as this gendering of the national privileges the masculine over the
feminine, so too does it privilege one particular notion of masculinity.
Nationalistic rhetoric is characteristically heterosexual/heterosexist, most
especially in its promotion of the nuclear family. The Eastern European
feminist writing that I read for this chapter, for example, is consistent in its
lack of concern for sexuality.

Thus far I have examined the scripting of the gendered subject into the
narrative of national community, but the constitution of nation is not only a
process which occurs within the confines of the nation itself: national identity
is also constructed through engagements with the international realm (see,

for example, Campbell 1992). The daily plebiscite of national identification constructs not only the national 'us' but also 'them', those who are outside and different. War memorials for example do not celebrate *the* victims of conflict but *our* dead. The nature of this construction of others is of fundamental importance to national identification (Johnson 1995).

The boundary separating the national from the international is thus also dependent upon gendered identities. Foreign territorial aspirations are usually described as being aggressive, but unnaturally so, a perversion of the 'normal' manner. Drawing on the traditions of Orientalism, Iraq was frequently evoked in gendered terms during the Gulf War, most blatantly in its 'rape of Kuwait'. Often perversity was defined from a homophobic position which accounts for the multiple plays in jokes and cartoons on the closeness of the name Saddam to the word sodomy (see Enloe 1993: Ch. 1). Also from the Orientalist tradition, the rhetoric and military practice of the Gulf War demonstrated a fear of an uncontrolled female sexuality, a sexuality which had to be disciplined and domesticated (Sparke 1994).

This discursive practice represents not only words but has significant material effect. Women have been systematically excluded from the international realm. Few women enter into the history of international relations, certainly after the age of European diplomacy by intermarriage of royal families has passed. The arena of international relations has been written as 'high politics':

> This supposedly autonomous realm of raison d'état is now staffed by national security managers, supposedly expert in the art of coordinating diplomacy, covert action, intelligence, and military preparation . . . [A]rguments from the Oliver Norths of this world inform us that this sphere of human activity is both beyond the comprehension of most Western publics and far too important for democratic oversight.
>
> (Dalby 1994: 534)

Yet the international system *is* dependent upon gender norms: by turning attention from key leaders, and refocusing upon the lives of those who are excluded from the formal sphere of high politics, it is possible to see the other side of the working of international political (and representational) economy. Enloe (1989, 1993) indeed goes as far as to say that international relations depend upon gender relations: the international division of labour *requires* the feminization of labour forces and the privatization of reproductive activities; tourism for example depends upon sensual mythologies of exotic places and cheap labour to service tourists.

CONCLUSION

Gender and nationality are significant elements of contemporary subject identity and yet neither gender nor nationality are a priori categories. Subjects are overdetermined by locations in multiple groups and processes. These multiple identifications are not additive; it is not possible to distil one aspect of identity such as gender or nationality for examination independent of other aspects (socioeconomic position, race, nationality and so forth). In other words, gender and national identities are, to an extent, constituted differently in each location. Even in this initial account of nation and gender for example, I have found it necessary to examine the construction of domesticity and international politics. By way of a conclusion I would like to pull out two further themes which have run through the chapter.

First, the unification of a people into nationhood is not a simple process. The creation of the appearance of unity is only possible through struggle, in the case described here, a contest between gendered identities. This is obviously not a simple case of men versus women but instead a recognition of the pressures and divisions which arise from employing 'gender to fashion a national community in somebody's, but not everybody's image' (Enloe 1993: 250).

Second, I hope to have illustrated the ambivalence of certain categories traditionally employed to discuss identity. The case of Eastern European feminisms has illustrated the ambivalence of the domestic: it is not only a space of containment of women, it is in certain places a site of resistance against state requirements for labour, and a space of radical politics aiming to transcend the status quo. The militarization of national citizenship has been exclusionary of women. Now, however, women are beginning to challenge the naturalization of male dominance of the military. While this will erode the masculinization of the military, providing women with more agency in this institution, it will be at the expense of any strength that feminism can attain by positing the masculinity of war, and feminism as an alternative citizenship based upon more peaceful characteristics usually labelled 'feminine'.

Out of this ambivalent political process there emerges a feminist analysis which is not solely concerned with constructing an identity of 'women *as* women' (Mouffe 1992: 382) but with mapping the complex societal relationships which construct dominance and subjugation: dominance not as a monolith but as overdetermined by a number of subjugations, one of which is centred around the construction of gender norms and differences. Feminism as I understand it then, is involved in this project of disentangling power and dominance, in denaturalizing and opposing the apparently 'natural' gender relations supporting of and supported by other forms of subjugation. Instead of trying to prove that any given form or scale of feminist discourse is the only one that reveals the 'real' essence of womanhood, one might better adopt a perspective that opens up possibilities for an understanding of the role of gendered identities in the construction of the multiple forms of subordination underwriting society.

ACKNOWLEDGEMENTS

I would like to thank Nancy Duncan for encouraging me to write this paper, and for her constructive comments on various drafts. Nirmala Ervelles, Doreen Massey and Paul Routledge also provided invaluable critiques of earlier drafts. Of course, all the usual disclaimers apply.

NOTES

1 Differential relations between men and women continue in the military, including sexual harassment. Women's access to the military does not therefore automatically guarantee emancipation (Yuval-Davis 1991: 64).
2 Communist imposition of women's 'liberation' onto East European societies distorts this politicization of identity.
3 Of course, going out of the home to find work is not a simple sense of liberation for poor women in the West either; here too it represents an economic necessity.
4 This sentiment should be understood in the context of material shortages endemic in communist and socialist countries of this century. Eastern European women often have difficulty accepting Western feminist critiques of consumerism as manipulating of women's identity because of the time they expend attempting to purchase limited commodities.

MASCULINITY, DUALISMS AND HIGH TECHNOLOGY

Doreen Massey

One important element in recent feminist analyses of gender has been the investigation and deconstruction of dualistic thinking. This chapter takes up one aspect of this issue of dualisms and the construction of gender. It examines the interplay between two particular dualisms in the context of daily life in and around high-technology industry in the Cambridge area of England. The focus on dualisms as *lived*, as an element of daily practice, is important (see Bourdieu 1977; Moore 1986). For philosophical frameworks do not 'only' exist as theoretical propositions or in the form of the written word. They are both reproduced and, at least potentially, struggled with and rebelled against, in the practice of living life. The focus here is on how particular dualisms may both support and problematize certain forms of social organization around British high-technology industry.

High-technology industry in various guises is seen across the political spectrum as the hope for the future of national, regional and local economies (Hall 1985) and it is important therefore to be aware of the societal relations, including those around gender, which it supports and encourages in its current form of organization.[1] In the United Kingdom 'high-tech' industry has been sought after by local areas across the country and has been the centre-piece of some of the most spectacular local economic success stories of recent years. In particular, it is the foundation of what has become known as the Cambridge Phenomenon (Segal Quince & Partners 1985). The investigation reported on here is of those highly qualified scientists and engineers, working in the private sector in a range of companies from the tiny to the multi-national, who form the core of this new growth. These are people primarily involved in research, and in the design of new products. This is the high-status end of high tech. The argument in this chapter takes off from two important facts about this part of the economy and the scientists who work within it: first, that the overwhelming majority of them are male; and second, that they work extremely long hours and on a basis which demands from them very high degrees of both temporal and spatial flexibility (see Henry and Massey, 1995). It was the conjunction of these two things which led to the train of enquiry reported here.

REASONS FOR THE LONG HOURS OF WORK

There are three bundles of reasons for the long hours worked by employees in these parts of the economy.[2]

The first bundle of reasons revolves around the nature of competition between companies in these high-technology activities. This is the kind of competition which has frequently been characterized as classically 'post-Fordist'. Production frequently takes place on a one-off basis, as the result of specifically negotiated, and competitive, tenders. High up among the criteria on which tenders are judged is the time within which the contract will be completed. Moreover, both during and after production there is a strong emphasis on responsiveness to the customer: in answering enquiries, in solving problems which emerge during and after installation/delivery of a product, in being there when needed – even if the telephone call comes through from California in the middle of the night. It is not so much the inherent unpredictability of R&D as the way in which it is compressed into the spatio-temporal dimensions required by this particular social construction of competition which is the issue. 'Time' is important to successful competition. The results should give pause for thought. For these are high-status core workers in what is frequently heralded as a promising flexible future. The demands which this flexibility places even on these workers are considerable.[3]

Moreover, these pressures for long hours are added to by a second bundle of reasons: those which revolve around the nature of competition within the labour market. There are a number of strands to this, but the most significant derives from the general character of this market as a knowledge-based labour market. It is a market in individualized labour power, valued for its specific learning, experience and knowledge. In order to compete in this labour market (and others like it) employees must, beyond the necessity of working the already-long hours required by their companies, continue to reproduce and enhance the value of their own labour power. They must keep up with the literature, go to conferences, and maintain the performance of networking, and of talking to the right people, and so forth. This is additional labour which is put in outside of the hours required by the company and its success, but equally necessary for the success of the individual employee. Within the workplaces, too, the interaction between employees can produce a culture which glorifies long hours of work. Again, this may derive from competition between individuals, but it may also result from various peer-group pressures – the need 'not to let the team down', for instance, can become a form of social compulsion (Halford and Savage 1995).

But the third reason that the employees in these parts of the economy work such very long hours is completely different. It is, quite simply, that they love their work. Figure 7.1 illustrates some aspects of this; the first four quotations are from scientists themselves, the last two from company representatives. These scientists and engineers become absorbed by their work, caught up by the interest of it; they don't like to leave an element of a problem unsolved before they break off for the evening. The way this is interpreted, or presented, by different groups varies. Thus, company representatives speak of the kinds

> We don't **need** to work longer – I think people choose to because they enjoy their work, because they own the project . . . and there's also ownership of the client.
>
> The clock doesn't matter at all. The only restriction for me is I don't like to get home too late. The landlady's given me a key, but I don't like to arrive much after midnight.
>
> I've got so much holiday I don't know what to do with it.
>
> . . . because I enjoy it . . . I enjoy the work . . . I enjoy computers . . . I often wonder what I would have done if I'd had to get a job in the days before computing.
>
> One person was sent abroad to a conference because they would not take time off.
>
> But the thing we have discovered over the years is that people who work here, and get into it, become addicted . . . we find the problem of getting some people to leave; they do get very engrossed in the thing . . . This circuit of people working on the system here, the difficulties are extracting them, for some other thing that may be necessary, like they haven't had any sleep for the last 40 years!

Figure 7.1 Enthusiasm for work leading to longer hours

of people they seek to employ as committed and flexible, as 'motivated', as 'able to take pressure', as not being the kind to watch the clock, and they not infrequently acknowledge that this characteristic may derive from pure interest in the work itself. A number of company representatives were quite clear that their search for employees was directed towards finding these characteristics. The scientists themselves often talk of their delight in the nature of the work, of its intrinsic interest. Where these male scientists have partners, however (and all the partners we identified were women), the partner's view of it was often more cynical, more tinged with the observation of obsessiveness, or of workaholism.[4]

A couple of points are worth making at this juncture. First, one's immediate response to the working lives of these employees may well be critical. Certainly, as we carried out the research, ours was so, in principle at least. Yet all of the reasons for this perhaps excessive duration of work also have their other side: each is thoroughly ambiguous though each in a different way. In terms of sectoral competition, 'putting the customer first' is no bad thing (especially if you are the customer). Yet the demands placed on employees can be enormous. In terms of the labour market, it has usually been interpreted as an advance that one's value is based on knowledge and experience rather than, for example, on lack of unionization, or the low level of the wages one is prepared to accept. Moreover, the individualization of the labour market must in some senses be an advance, certainly over the treatment of workers as a mass, as an undifferentiated pool of nameless labour power. The idea that we are heading towards an economy and society which is based on knowledge, however unlikely in fact, has always been treated as a change for the better. Finally, the fact that people enjoy their work, and that they enjoy it in part precisely because it is knowledge-producing (in the employees themselves) can only be seen as an improvement over the kinds of jobs which are characterized above all by mind-numbing monotony and a desire to get to the end of the day. After years of exposing the fact and the effects of deskilling I find it hard to criticize jobs because they are too absorbing and demand too much in the way of skill-enhancement! (Yet this very dilemma may point to the fact that the problem has been wrongly posed. Maybe it is the polarization between deskilled and super-skilled which should be the focus of our attention . . . ?)

A second point worth noting is that the second and third of these reasons for long hours (the nature of the labour market and the love of the work), though perhaps less so in its particular articulation the first (the nature of competition in the sector), are shared by many other occupations and parts of the economy, especially professional sectors and perhaps most particularly academe. Some of the issues which arise are therefore of much more general relevance, beyond the relatively small sectors of high technology in Cambridge. Certainly, they posed questions to us personally, as we did the research. Yet in other ways, the particular manner in which these pressures function, and the kinds of social characteristics with which they are associated, are quite specific to individual parts of the economy.

DUALISMS AND MASCULINITIES

One of the specificities of these high-technology sectors is bound up with the *reasons why* the employees are so attached to their jobs and how these are interpreted. The dynamics in play here are bound up with elements of masculinity, *and of a very specific form of masculinity*. Above all, the attachment to these jobs is bound up with their character as scientific, as being dependent upon (and, perhaps equally importantly, confined to) the exercise of rationality and of logic. Within the structure of the economy, these jobs represent an apex of the domination of reason and science. It is this which lends them much of their status and which in part accounts for the triumphalist descriptions they are so often accorded in journalistic accounts. What is demanded here is the ability to think logically.[5] It is, in other words, a sector of the economy whose prime characteristics, for these employees, are structured around one of the oldest dualisms in Western thought – that between Reason and non-Reason; and it is identified with that pole – Reason – which has been socially constructed, and validated, as masculine (see, especially, Lloyd 1984).

Science, moreover, in this dualistic formulation is seen as being on the side of History (capital H) as progression. It makes breakthroughs; it is involved in change, in progress. And it is here that it links up to the second dualism which emerged as this research proceeded: that between transcendence and immanence. In its aspect of transcendence, science is deeply opposed to that supposed opposite, the static realm of living-in-the-present, of simple reproduction, which has been termed immanence. This opposition between transcendence and immanence is also a dualism with a long history in Western thought. And again it is transcendence which has been identified as masculine (he who goes out and makes history) as against a feminine who 'merely' lives and reproduces. As Lloyd (1984: 101) argues:

'Transcendence', in its origins, is a transcendence *of* the feminine. In its Hegelian version, it is associated with a repudiation of what is supposedly signified by the female body, the 'holes' and 'slime' which threaten to engulf free subjecthood (see Sartre, 1943, pp. 613–14) . . . In both cases, of course, it is only from a male perspective that the feminine can be seen as what must be transcended. But the male perspective has left its marks on the very concepts of 'transcendence' and 'immanence'.

The two dualisms (Reason : non-Reason, and Transcendence : Immanence) are thus not the same, though there are interrelations between them.

The reasons for these characterizations, and for the construction of these dichotomies in the first place, and their relationship to gender, have been much investigated (see, for instance, Dinnerstein 1987; Easlea 1981; Hartsock 1985; Keller 1982, 1985; O'Brien, 1981; Wajcman 1991). Many of these authors have examined the relation between the constitution of science on the one hand and of gender on the other. David Noble's (1992) history of 'a world without women' tells the long story of the capturing by enclosed masculine societies of the kind of knowledge production which was to receive the highest social valuation.

Such dualist thinking has, as has already been said, been subject to much criticism. However, the nature of the criticism has changed and been disputed. In *The Second Sex* (1949/1972) Simone de Beauvoir famously urged women to enter the sphere of transcendence. In recent years, however, it has rather been the fact of thinking dualistically which has been objected to. Dualistic thinking has been criticized both in general as a mode of conceptualizing the world and in particular in its relation to gender and sexual politics. In general terms, dualistic thinking leads to the closing-off of options, and to the structuring of the world in terms of either/or. In relation to gender and sexuality it leads, likewise, to the construction of heterosexual opposites and to the reduction of genders and sexualities to two counterposed possibilities. Moreover, even when at first sight they may seem to have little to do with gender, a wide range of such dualisms are in fact thoroughly imbued with gender connotations, one side being socially characterized as masculine, the other as feminine, and the former being accordingly socially valorized. The power of these connotational structures is immense, and it is apparently not much lessened – indeed it is possibly only rendered more flexible – by the existence among them of inconsistencies and contradictions.

It was only gradually, in the course of considering the interview material and the nature of work in the scientific sectors of the economy, that the issue of dualisms emerged as significant in this research. It was the things which people said, the way life was organized and conceptualized, the unspoken assumptions which repeatedly emerged, which pushed the enquiry in this direction.

Thus, for example, it was evident that in Cambridge these scientific employees were specifically attached to those aspects of their work which embody 'Reason' and 'Transcendence'. What they really enjoy is its logical and scientific nature. They themselves when talking may glory in the scientificity of their work, and frequently exhibit delight in the puzzle-solving logical-game nature of it all.[6] Their partners comment upon their obsession with their computers, and both partners and company representatives talk of boys with toys (one representative candidly pointing out that these guys like their jobs because the company can buy far more expensive toys than the men themselves could ever afford):

> 'We have toys which they can't afford. You know engineers, big kids really; buy them a computer, you know you've got them . . . you know [they are] quite happy if you can give them the toys to play with.'

This attachment to computers may be seen in this context as reflecting two rather different things, both of which are distinct from the more technologically oriented love of 'fiddling about with machines'. On the one hand these machines, and what can be done with them, embody the science in which the employees are involved. They are aids and stimuli to logical thought. On the other hand their relative predictability (and thus controllability) as machines insulates them from the uncertainties, and possibly the emotional demands, of the social sphere.

The aspect of transcendence comes through in the characterizations of the job as 'struggling' with problems, as 'making breakthroughs'; whether they think of themselves as far from it or right up against it there is the notion of a scientific-technical 'frontier'. One scientist, reflecting on the reasons for his long hours of work, talked of being 'driven by success' and the fact that he was 'always reaching higher'. Another scientist in the same company, but who was quite critical of the hours worked by others, argued that for some people crisis is part of the job culture: 'it's a sort of badge of courage'. Other words, too, reflect the effort and the struggle of it all: 'If I stagger out of here at 11 o'clock at night I really don't feel like going home and cooking'. There's the quest: 'As a parent I try to spend as much time as I can with [the child] but in my quest for whatever it is I tend to work very hard'. There's the compulsion: 'if you've gotta do it then you've gotta do it'. And, hopefully, there's the triumph:

> 'his wife is much more even-tempered than my wife who says sort of "What the hell, Friday night we have got to go out and don't you forget it", but [my wife] accepts the fact that if there is nothing on specifically and nothing to be done, that the chances are that I will disappear, and reappear looking cross-eyed and what not, with a slightly triumphant smile or look downcast.'

That quotation illustrates also a further phenomenon: that the self-conception of many of these employees is built around this work that they do and around this work specifically as scientific activity:

'the machine in front of them is their home . . . '

'It is their science which dominates their lives and interests . . . '

Moreover, this glorification of their scientific/research and development capabilities on the part of the scientists can go along with a quite contrasting deprecation of their ability to do other things, especially (in the context of our interviews) their incompetence in the face of domestic labour. This is work which it is quite acceptable *not* to be good at. Thus:

> *laundry?* 'I shove it in the machine'; *cleaning?* 'I do it when it gets too much'; *shopping?* 'Tescos, Friday or Saturday'; *cooking?* 'I put something in the microwave. Nothing special. As long as it's quick and easy that's good enough for me'; *gardening then?* 'when necessary'.

There is here none of the pleasurable elaboration on the nature of the tasks which typifies descriptions of the paid scientific work. The answers are short and dismissive.

Such attitudes are important in indicating what is considered acceptable as part of this scientist's own presentation of himself. Not only is the identification with scientific research very strong and positive, but it seems equally important for him to establish what is *not* part of his picture of himself. Domestic labour and caring for his daily needs and living environment is definitely out. It is not just that scientific activity is positively rated which is significant but the fact that it is sharply cut off from other aspects of life. This is precisely the old dualism showing its head in personal self-identification and daily life. What was going on was a real rejection of the possibility of being good at *both* science *and* domestic labour. A framing of life in terms of 'either/or'.

In this case, and in some others, such downplaying of the rest of life extended to all non-work/scientific activities. But such extreme positions were not common and seem to be more evident among single men than those with partners, and, even more markedly, than among those with children. Some men were clearly aware of the issue. For one scientist, a new baby had 'completely changed his life' (what this meant was that he went home early almost every other night), and yet the difficulty of balancing or integrating the sides of life was evident:

> 'I feel frustrated . . . when . . . after this baby that's changed my life . . . I go home early every other day (almost) and pick her up at 4.35, take her home, play with her until bedtime, and . . . I find that sometimes that's quite frustrating, and keeps me away from work. I mean – it's fulfilling in its own right, but it's . . . I'm conscious of the fact that . . . I call it a half-day, you know. I find it frustrating.'

Finally, some of the comments made about the scientists by (some of) the partners were particularly sharp and revealing:

They're

' . . . not very socially adequate.'

' . . . better with things than with people.'

'Work gets the best of him.'
'Work is the centre of his life.'

One of the very few female company representatives (that is, a member of management, not of the scientific team) reflected: 'Well, when I first joined the company there were twelve people here and they stuck me in an office with the development team and it was a nightmare. I really hated it. They didn't talk, they didn't know how to talk to a woman, they really didn't.'

What appears to be going on, in and around these jobs, is the construction/reinforcement of a particular kind of masculinity (that is, of characteristics which are socially coded masculine) around reason and scientificity, abstract thought and transcendence. It is a process which relates to some of the dualisms of Western thought and which, as we shall see further, has concrete effects in people's lives.

Such characteristics of the employees, it must be stressed, relate to the more general nature of these jobs. These are jobs which derive their prestige precisely from their abstract and theoretical nature. They are jobs the very construction and content of which are the result of a long process of separation of conception from execution (and of the further reinforcement of this distinction through social and spatial distancing). They are jobs, in other words, which enable and encourage the flourishing of these kinds of social characteristics. Moreover the long hours which, for the various reasons discussed above, are worked in them enforce both their centrality within the employees' lives and a passing on of the bulk of the work of reproduction to others. In Cynthia Cockburn's words: 'Family commitments must come second. Such work is clearly predicated on not having responsibility for child-care, indeed on having no one to look after, and ideally someone to look after you' (Cockburn 1985: 181). The implication of all this is not only that these jobs are an embodiment in working life of science and transcendence, but also that in their very construction and the importance in life which they thereby come to attain, they enforce a separation of these things from other possible sides of life (the Other sides of Reason and of Transcendence) and thus embody these characteristics *as part of a dualism*. Moreover by expelling the other poles of these dualisms into the peripheral margins of life, and frequently on to other people (whether unpaid partner or paid services), they establish the dualisms as a social division of labour. The pressure is for someone else to carry the other side of life.

Moreover, if there is indeed a form of masculinity bound up with all this, then the companies in these parts of the economy let it have its head; they trade on it and benefit from it, and – most significantly from the point of view of the argument in this chapter – they thereby reinforce it. Furthermore, the possession of these characteristics, which are socially coded masculine and which are related to *forms* of codification which resonate with dichotomous distinctions between two genders, makes people more easily exploitable by these forms of capital. There is here a convergence of desires/interests between a certain sort of masculinity and a certain sort of capital.[7]

This is not to say that what is at issue here is simple 'sexism'. Our interviews – certainly as analysed so far – did not reveal the explicit sexism found in some other studies, including Cockburn's (1985). We did not encounter much in the way of strong statements about the unsuitability of women for these jobs. There were a few such statements but they were infrequent in the overall context of our interviews. Nor was it clear that the male scientists who displayed the characteristics which have been described always recognized them explicitly as masculine (although further probing may well have unearthed more evidence on this score). The point, however, is that what is at issue here is not so much overt discrimination or sexism as deeply internalized dualisms which structure personal identities and daily lives, which have effects upon the lives of others through structuring the operation of social relations and social dynamics, and which derive their masculine/feminine coding from the deep socio-philosophical underpinnings of Western society.

THE WORK/HOME BOUNDARY

The boundary between work and 'home' has often been seen, and in this case can be seen, as an instantiation of the dualism between transcendence and immanence.[8] At work the frontiers of history are pushed forward; at home (or so the formulation would have us believe) there is a world of feelings, emotions and (simple) reproduction. Lloyd (1984: 50) once again, summarizes the complex arguments which have evolved:

> We owe to Descartes an influential and pervasive theory of mind, which provides support for a powerful version of the sexual division of mental labour. Women have been assigned responsibility for that realm of the sensuous which the Cartesian Man of Reason must transcend, if he is to have true knowledge of things. He must move on to the exercise of disciplined imagination, in most of scientific activity; and to the rigours of pure intellect, if he would grasp the ultimate foundations of science. Woman's task is to preserve the sphere of the intermingling of mind and body, to which the Man of Reason will repair for solace, warmth and relaxation. If he is to exercise the most exalted form of Reason, he must leave soft emotions and sensuousness behind; woman will keep them intact for him.

The fact that all this can be, and has been, severely criticized in terms simply of its descriptive accuracy, most particularly from a feminist perspective, has not destroyed its power as a connotational system. What is at issue in the ideological power of these dualisms is not only the material facts to which they (often only very imperfectly) relate (many women don't like housework either, and many female paid employees negotiate a work : home boundary), but the complex connotational systems to which they refer. Moreover, the negotiation of this boundary has emerged in our research as a crucial element in the construction of these men's attitude to their work, and in their construction of themselves.

One of the avenues of enquiry which originally sparked my interest in designing this research derived from statements made in interviews in a previous project (Massey *et al.* 1992). That project also was concerned with investigating high-tech firms, in this case specifically those located on science parks, and one of the recurring themes in a number of the interviews concerned the blurring of boundaries. 'The boundary between work and play disappears' was a formulation which stuck in my mind. What absorbed me at that point was the characterization of everything outside of paid work as 'play' and, especially given the very long hours worked in the companies we were investigating, it prompted me to wonder who it was that performed the domestic labour which was necessary to keep these guys fed and watered and able to turn up for work each morning. (The work of 'domestic labour', who performs it and how, and the complex intra-household negotiations over it, is the subject of another forthcoming paper.) But what the interviewee had in mind was the fact that work itself had many of the characteristics of play: that you get paid for doing things you enjoy, you have flexible working arrangements, you take work home, you are provided with expensive toys. In this formulation, there really is no boundary between paid work and play. In this way of understanding things, 'the home' in the sense of the domestic, of reproduction, of the sphere of emotions, sensuality and feelings, or of immanence, does not enter the picture at all.

How then do we interpret what actually happens to the boundary between work and home in the case of these scientists in Cambridge? There are two stages to the argument.

First, there is indeed a dislocation of the boundary between work and home. Most particularly, this is true in a temporal and spatial sense. Moreover it is a dislocation which primarily takes the form of an invasion of the space and time of one sphere (the home) by the priorities and preoccupations of the other (paid work). This can be illustrated in a whole range of ways. The high degree of temporal flexibility in terms of numbers of hours worked turns out in practice to be a flexibility far more in one direction than in the other. While the demands, and attractions, of work are responded to by working evenings, weekends, Bank Holidays and so forth – and it is expected that this will be so, it is the 'commitment', and 'flexibility' required to be an accepted member of this part of the economy – the 'time-in-lieu' thereby in principle accrued is far less often taken and indeed has to be more formally negotiated, and the demands of home intrude into work far less than vice versa (see Henry and Massey, 1995). Or again, spatially, work is very frequently taken home. A high proportion of these employees have machines, modems and/or studies, in the space of the domestic sphere, but there is no equivalent presence of the concerns of home within the central space of paid work (at the most obvious level, for example, not one of the companies we investigated had a creche). One of the company representatives we interviewed spoke of the employees being 'virtually here' (in the workplace) even when working at home, because of the telecommunications links installed between the two places. Moreover this raises a third and very significant aspect of this one-way invasion. A lot of our interviewees spoke of the scientists' difficulty in turning off thoughts about work, of not thinking about the problem you are puzzling over, even when physically doing something quite different. The men wondered if they should charge to the company time spent thinking in the bath. A few of both men and partners spoke of episodes when he would get up in the middle of the night to go and fiddle with some puzzle. Men, partners and sometimes children, commented on minds being elsewhere while officially this was time for playing with the children or driving the car on a day out. Here there is a real 'spatial' split between mind and body. Here there really is a capsule of 'virtual' time–space of work within the material place of the home. While the body performs the rituals of the domestic sphere the mind is preoccupied with the interests and worries of work.

'I am well aware of the fact that in many areas, that you are better having the 9–5 pm and everything like that, but I have never found it at all compatible with trying to work or trying to pursue a bit of research or a bit of development, to have to give up at the magic hour or whatever . . . and I mean you can't say to somebody you will think between 9 and 5 pm and you will not think between 5.05 pm and 8.55 am.'

This is eminently understandable, and in many ways an attractive situation – it is good to have paid employment which is interesting, and it is a challenge to resist the compartmentalization of life into mutually sealed-off time–spaces. But what is important is that once again this only works one way. While domestic time is in this sense porous, work time is not. Indeed, and this is the significant point, it *cannot* be so. While it is assumed that one may think about work while playing with the children or while out for the day with the partner, the reverse is not the case. Indeed a reason quite frequently given for working late nights and weekends at the office is that the time–space is less disturbed then – even if other people are doing the same thing there is less in the way of incoming 'phone-calls and so forth. One of the dominant characteristics of this kind of work is that it demands, and induces, total concentration. The above quotation is interesting in its implication that 'thought' is involved only in paid work.[9] Moreover it is the kind of thought which requires a lack of intrusion; it is totally absorbing. Even the reservations to this 'all-work' atmosphere of the workplace in a sense reinforce its truth. Thus one or two workplaces had a gym and elaborate catering facilities on site, the aim being to aid rather than detract from the overall ability to concentrate. And in one company, partners – seemingly in despair at ever seeing their men – came into the office:

' . . . they have children and wives and they are always retailing the complaints from their wives . . . '
'This is a constant complaint . . . there is a perennial complaint that the partner never sees them and they are always in here. In fact, partners tend to come in here and work in the evenings because that's where the other one is; they have different kinds of jobs but they can bring their work with them and do it here.'

In fact, it is hardly an invasion: she is conforming to the norms of the workplace; what she has brought in with her is her 'work', not the sphere of the domestic, and he can carry on with what he has to do.
This does not mean that levels of concentration within the workplace do

not vary, nor that time-out cannot be taken. Indeed time-in-lieu, trips to the shops, etc., provide occasional windows within the working day. But within the workplace, everything, even the exercise of the body, is geared to the productivity of the intellect:

'I was amazed when I went there – I'd been working at [major corporation]. This huge factory in Lancashire had shut and I came down here to the interview with [smaller company, Cambridge-based] and I walked up the stairway and on every floor there was a series of little offices and ramps around the edge and the middle of each floor was open and there was a ping-pong table or a snooker table and everybody seemed to be playing games and I thought that this is supposed to be a place of work – and then when I saw all the things they were doing – a chap put his bat down and [would] go off and design an IC in a little room in the corner.'

What we have here, then, is the workplace constructed as a highly specialized envelope of space–time, into which the intrusion of other activities and interests is unwanted and limited.[10] 'The home', however, for most of these scientists, is constructed entirely differently. Both temporally and spatially it is porous, and in particular it is invaded by the sphere of paid work.

One way of beginning to conceptualize the difference between these two kinds of spaces is through the work of Henri Lefebvre. In his account of *The Production of Space* (1991), he characterizes the space of current Western society as 'abstract space' and discusses (and criticizes) as one of its defining features its fragmentation, its division into sub-spaces devoted to the performance of specialized activities. His historical analysis explains this process as the result of aspects both of modernity and capitalism on the one hand and of currently dominant forms of masculinity on the other. Although Lefebvre's historical account, and the supposed newness of abstract spaces, may be questioned, his examples of such specialized and fragmented spaces/space–times resemble very strongly the specialized space–times constructed in high-tech workplaces. They seem to have many of the characteristics of abstract space: they are demarcated against an outside, they are specialized, they are masculine. Yet in the story we are telling here they are not coexisting with other similarly specialized and sealed-off time–spaces but with a time–space, that of the domestic sphere, which is porous, which allows entry from other spheres, which is perhaps in Lefebvre's terms characteristic of an older, and yet possibly at the same time more potentially progressive, kind of time–space. Lloyd, it might be recalled, contrasted the wholly rational sphere of Reason/ Transcendence (i.e. evacuated of other things) with 'woman's task' of preserving 'the sphere of the *intermingling* of mind and body' (1984: 50; my emphasis).

Further, Lefebvre pointedly asks

Is not social space always, and simultaneously, both a *field of action* (offering its extension to the deployment of projects and practical intentions) and a *basis of action* (a set of places whence energies derive and whither energies are directed)?

(1991: 191)

In other words, social space is both an arena of action and potentially enabling/productive of further effects. Just so the places of work in these high-tech parts of the economy: they are not merely spaces where things may happen but spaces which in the nature of their construction (as specialized, as closed-off from intrusion, and in the nature of the things in which they are specialized) themselves have effects – in the structuring of the daily lives and the identities of the scientists who work within them. Most particularly, in their boundedness and in their dedication to abstract thought to the exclusion of other things, these workplaces both reflect and provide a material basis for the particular form of masculinity which hegemonizes this form of employment. Not only the nature of the work and the culture of the workplace but also the construction of the space of work itself, therefore, contributes to the moulding and reinforcement of this masculinity (see also Massey and Henry, forthcoming). As Lefebvre writes: 'The dominant tendency fragments space and cuts it up into pieces . . . Specializations divide space among them and act upon its truncated parts, setting up mental barriers and practico-social frontiers' (1991: 89). Lefebvre would argue that the currently dominant tendency towards the homogenization/fragmentation and specialization of space is something which should be opposed. This relates to the second stage in the argument here about what is happening to the work/home boundary among the scientists of the Cambridge phenomenon.

For what has been discussed so far is an alteration in the boundary between home and work which consists of nothing more than the spatio-temporal transgression by one sphere (one side of the dualism) into the other. The first point to have been noted is that this transgression is all one way. The second stage of the argument, however, is that in whatever manner one interprets this 'blurring' of boundaries it does not entail any kind of overcoming of the dualism itself. Yet it is the fact of dichotomy itself (Reason/non-Reason; Transcendence/Immanence) which has been criticized as being part of that same mode of thinking which also polarizes genders and the characteristics so frequently ascribed to them. And it is the parallel fragmentation/specialization which came in for criticism from Lefebvre. What, then, can be learned about the possibility of unification from this study of Cambridge scientists?

RESISTANCE

The characteristics which have been described above are traits of *masculinity*, not of men. As already implied there is no simple homogeneity among the men we studied. However, these characteristics are strongly embedded within the culture of this part of the economy (with some variation in detail between different types of jobs). Moreover, the strength of this embeddedness means that these characteristics 'pull' all its participants towards them. Individual men have relations to these characteristics which are more or less celebratory or painful. Many of them recognize the need to negotiate the very different personas they inhabit at home and at work – the scientist with the new baby (quoted earlier) was doing just that. And what he was confronting there was precisely the difficulty of preventing his dominant self-conception as a scientist from completely overriding those other potential sides of himself. Other men actively try to resist his potential domination. Their number is small and their reasons varied. Most commonly resistance is a response to stress or to strongly articulated objections on the part of the partner, or to a genuine sensitivity to the men's felt need to live a more varied life, not to miss out on the children growing up, and so forth.

Moreover, the resistance takes a particular form. It is almost entirely to do with working hours, and with the time and space which work occupies, rather than with wider characteristics of the job. It also takes place almost entirely at the individual level. These workplaces are not unionized. Moreover, at a more general social level, while there are trade-union campaigns and feminist arguments for a shorter working day and week, they have as yet made very little progress. Certainly there seems to have been no thoroughgoing cultural shift, in spite of the increasing proportion of employment which is part time, in favour of shorter working hours. Indeed, since in these parts of the economy at least some of the compulsion to work long days comes from the interest in and commitment to the work itself it is not clear how such jobs and others like them relate to the wider arguments about working time.

Given all this, it is the scientists individually who decide how they are going to respond to the pressures and attractions of their jobs, and how they will negotiate the work/home boundary and the different identities they may imply.

In this context it is deeply ironic that one of the important mechanisms of resistance, and one adopted by a number of the men, is precisely to insist on the necessity for and the impermeability of the boundary between work and home. Given the fact that the tendency is for work to invade home life one obvious mechanism for resistance is to protect home life from intrusion. This happens in a number of ways. Some men (a few only, but then the resisters in total are not a high proportion of the whole) have decided not to take work home, thereby preserving the space of home and the time spent in it from the intrusion of the demands of paid work. Sometimes this will involve an intrusion in time terms, maybe involving staying longer at the workplace in order to finish a task there rather than take it home. It is here the *space* of home which is seen as being the most important not to violate. Other men, though again only a few, have made themselves rules about *time* and insist on keeping to a regular daily routine and on arriving and leaving the workplace at set times. Over the long term it is possible that this will be detrimental to their careers (see Henry and Massey, 1995), but the men are aware of this and indeed in some cases have adopted the strategy because of other problems (personal stress, problems with health or personal relationships) which had been produced by a previous commitment to the high pressure and long hours more typical of these companies in general. It must be emphasized that this is not the only way of coping with the pressures of this work where they are experienced as a problem. Other scientists, and couples, have found other ways of dealing with the demands and compulsions of this kind of work but what is significant about this one is its irony. The 'problem', as we have argued above, has been posed through the working out in everyday life of some of the major dualisms of Western ways of thinking. Yet, in the absence of collective resistance, legislative action or wider cultural shifts, individual attempts to deal with some of the conflicts thus provoked may result in a reinforcement of the expression of those very dualisms. The dichotomies are rigidified in order to protect one sphere (the home, the 'rest of life') from invasion by the other (scientific abstraction, transcendence). The problems posed by the dualisms result in their reinforcement.

CONCLUSIONS

The last section concluded on an irony: that those who were attempting to resist the domination of their lives by one side of a dualistic separation most often found themselves reinforcing the divide between the two poles of the dualism. This was one among a number of ironies in the situation analysed in the chapter. What such Catch-22 situations indicate is that the way out of the conundrums does not lie at that level. The 'solution' must be sought in a deeper challenge to the situation.

Similarly, the empirical material discussed here raises a number of confusions and complexities around the politics of campaigns for a shorter working day/week. They are issues, too, which relate as much to academe, especially in its present increasingly intensified and individually competitive form, as they do to the high-tech work discussed in the chapter. They are issues which touched me personally as an academic and which made me think about my own life as I did the research. It is a privilege to have work which we find interesting. At a recent meeting of feminist academics, where we discussed an early version of this paper, *none* of us wanted our 'work' to be restricted to 35 specified hours in each week. While all of us wanted to resist the current pressures on hours produced by the reinforcement of competitive structures, we did not want to lose either the feeling of autonomous commitment or the possibility of temporal flexibility. But neither did we like the actual way in which this 'flexibility' currently works – the pressure towards what can only be called a competitive workaholism and the inability to keep things under control. These are things which we as academics, as well as those in the high-technology sectors discussed here, need to confront. For when an important element of the pressure on time results from personal commitment on the one hand and individualized competition on the other, as well as from sectoral and workplace cultures, how can any form of collective resistance be organized?

In the longer term the aim must be to push the questioning further, to try to find those solutions which may exist at 'deeper' levels. In particular, I suggest, it means *questioning the dualisms themselves*. That is, instead of endlessly trying to juggle incompatibilities, and to resolve ambiguities which in reality point to contradictions, it is important to undermine and disrupt the polarizations which are producing the problem in the first place.

In philosophy, and in particular in feminist critical philosophy, this position is by now well established. The aim in general is not now only to valorize the previously deprioritized pole of a dualism (as Simone de Beauvoir did) but to undermine the dualistic structure altogether.

Such more fundamental critiques may be carried into other areas. Thus, in the early part of this chapter I wrote of the difficulty I had found, after years of criticizing deskilling within industry, in finding myself criticizing jobs for being too absorbing. This was another irony indeed! However, as was hinted there, it may be that the very dilemma points to the fact that the issue would be better posed in another way. Rather than being critical of deskilling or super-skilling as such, it is the polarization between them which should be the focus of critical attention. What is at issue here – and it is an issue which again involves us as academics – is the *social division* between conception and execution, between intellectuals and the rest.

What I find more problematical as a political issue is the division of the lives of the scientists described in this chapter between abstract and completely 'mental' labour, on the one hand, and the 'rest of life' on the other. In the version of this chapter which was sent to referees I had unreservedly applauded those few attempts which we had come across to resist the compartmentaliza-tion of life into mutually sealed-off time–spaces. At least one referee questioned this, asking simply '*Why* is it good to resist compartmentalization?' And I know for myself that one thing I thoroughly enjoy is to sit down in the secluded and excluding space of the Reading Room at the British Museum and devote myself entirely to thinking and writing. And yet . . . to return to Lefebvre, do we want lives sectioned-off into compartments, into exclusive time–spaces: for the intellect, for leisure, for shopping . . . ?

This dilemma might relate to, and be partially addressed by, considering the major dualism discussed in this chapter – that in which 'Science' itself is involved. It is perhaps that the problem lies most fundamentally in the postulated separation-off of the isolated intellect from the rest of one's being, and calling the product of the working of that (supposedly) isolated intellect 'knowledge'. Among many others Ho (1993: 168) has argued for an alter-native:

> This manner of knowing – with one's entire being, rather than just the isolated intellect – is foreign to the scientific tradition of the west. But . . .
> it is the only authentic way of knowing, if we [are] to follow to logical conclusion the implications of the development of western scientific ideas since the beginning of the present century. We have come full circle to validating the participatory framework that is universal to all indigenous knowledge systems the world over. I find this very agreeable and quite exciting.

The real irony, then, may be that the long-standing Western (though not only Western) dualism between abstract thought and materiality/the body may lead through its own logic to its own undermining. And it is on that dualism that much of the separation within the economy between conception and execution – and thus these 'high-tech' jobs themselves – has been founded.

ACKNOWLEDGEMENTS

I would particularly like to thank Nick Henry, with whom much of the empirical work for this paper was done, for much discussion and comment. I also had the benefit of four extremely thoughtful sets of referees' comments – many thanks. A first version of this paper was presented in a seminar series at Syracuse University. I should like to thank Nancy Duncan for her invitation. The paper has been published in *Transactions of the Institute of British Geographers*, NS 20: 487–99, 1995.

NOTES

1 Only one aspect of these relations is explored in this chapter. The work forms part of a wider project on high technology and the social relations which surround it. This research was funded by the ESRC: R000233004: 'High-status growth? Aspects of home and work around high-technology sectors' and is being carried out with Nick Henry, now at the Department of Geography, University of Birmingham.

   The project forms part of a wider programme of five pieces of research on the nature and consequences of growth in the South-East of England in the 1980s. The programme is based in the Geography Discipline, in the Faculty of Social Sciences, The Open University, from where further information, and a series of Occasional Papers, are available.

2 The first two of these reasons are explored in more detail in Henry and Massey (1995). As part of the research we interviewed representatives of 19 companies, 60 male scientists and 38 partners, all of whom were female. 'Partnership' was defined in terms of cohabitation. About one-third of the scientists were not cohabiting. The quotations from interviews which are cited in this paper have been selected as *symptomatic*. They capture, or express with precision, points or attitudes which were typical or widely prevalent or, if indicated so in the text, which characterized attitudes held by some among the interviewees.

3 In a paper currently in production this characterization of the work as 'flexible' is itself being questioned (see Massey and Henry, forthcoming). Thanks to one of the referees for extended and constructive thoughts on aspects of this issue.

4 One result of this absorption in their work is of course that these men have less time over than they otherwise might for life in the domestic sphere. A future paper deals directly with this issue. In discussions on the present chapter, Cynthia Cockburn wondered

   whether the time stolen by these men to sustain their addictive habit may actually not be stolen from the home (other men don't spend more time than they have to in the home), but rather stolen from pub, club and trade union.
   (personal communication)

   There is probably a lot in this. The point in the present chapter is precisely to emphasize that what characterizes these sectors is a *particular form of* masculinity.

5 Cynthia Cockburn has pointed to some of the inconsistencies and contradictions even here – see her treatment of the concept of 'intuition', and of the scientists' ambiguous relation to it, in Cockburn (1985). Indeed, the very fact that the men 'really love' their work, are 'obsessive' and so forth touches on realms outside that of pure Reason. But as pointed out in the opening paragraph, consistency has never been the outstanding attribute of the functioning of these dualisms, nor has *inconsistency* seemed much impediment to their social power.

6 Similar worlds have been described by Tracey Kidder (1982) and Sherry Turkle (1984).

7 These interconnections between gender analysis and aspects of economic growth, and specifically economic geography, are explored further in a forthcoming paper.

8 While the home/work distinction may validly be read as an instantiation of this dichotomy it must be stressed that there is far more to the possibilities of 'immanence' than having children and doing the housework.

9 This view was reinforced in some cases by the contrast in attitude to the skills of paid work on the one hand and the domestic sphere on the other. Thus a number of the scientists ascribed the fact of their partner doing almost all of the housework to the fact that 'she's better at it'. The interesting thing here is that there seems to be no understanding that this skill is one which could be learned. In contrast to the highly intellectual paid jobs, for which much learning was necessary, this skill seems to be seen, although *implicitly*, as innate.

10 This is broadly true of most workplaces, though to different degrees. The windowless boxes of so many modern factories precisely demonstrate the desire not to let the eye/mind wander 'outside' during working hours. But in the kinds of employment under discussion here, together with some others, it is especially marked.

RENEGOTIATING GENDER AND SEXUALITY IN PUBLIC AND PRIVATE SPACES

Nancy Duncan

INTRODUCTION

In this chapter I offer a general introduction to the issues of regulating and negotiating gender and sexuality through the opposition of the public and the private. I argue that the binary distinction between private and public spaces and the relation of this to private and public spheres is highly problematic. Although it is a distinction encoded in law and deeply rooted in North American and British cultures, it is nevertheless unstable and often problematically conflated with related distinctions such as that between domestic or familial autonomy and public spheres. Increasing privatization, commercialization and aestheticization of public space has tended to depoliticize space and shrink public spheres. However, I will discuss various ways that the spatial and political practices of marginalized groups such as abused women and sexual minorities (lesbians, gays and sex workers) work to undermine the (always already unstable) coherence of this binary and related binaries. The destabilizing of this boundary is a countervailing force working to open up not only private space but to reopen public space to public debate and contestation.

One could choose other groups such as the homeless[1] with an interest in transgressing the public/private dichotomy. However, I have chosen abused women and sexual minorities because members of such marginalized groups have experienced acute spatial dissonance and in some cases have found workable strategies for resisting the spatial framework and dominant spatial practices of Anglo-American society. I will also discuss various spatial practices that work to reinforce this boundary and some of the tensions surrounding the concept of privacy implied by the boundary. By pointing to examples drawn from these marginalized groups I attempt to show some of the complexities and subtleties of oppression on the basis of spatially constituted gender and sexuality. I then conclude with a discussion of the need to further unpack and destabilize this binary distinction. My focus will be on contemporary North America and Britain.

THE PUBLIC AND THE PRIVATE

The distinction between the public and the private is deeply rooted in political philosophy, law, popular discourse and recurrent spatial structuring practices.

These practices demarcate and isolate a private sphere of domestic, embodied activity from an allegedly disembodied political sphere that is predominantly located in public space. The public/private dichotomy (both the political and spatial dimensions) is frequently employed to construct, control, discipline, confine, exclude and suppress gender and sexual difference preserving traditional patriarchal and heterosexist power structures.[2] Although the social and political problems to which I refer clearly have spatial (material, corporeal) components, the solutions to these problems will by no means be purely spatial or environmental ones. There is no question, however, that confinement (voluntary and forced) in private spaces contributes to a reduction in the vitality of the public sphere as a political site and diminishes the ability of marginalized groups to claim a share in power.

It is clear that the public–private distinction is gendered. This binary opposition is employed to legitimate oppression and dependence on the basis of gender; it has also been used to regulate sexuality. The private *as an ideal type* has traditionally been associated and conflated with: the domestic, the embodied, the natural, the family, property, the 'shadowy interior of the household', personal life, intimacy, passion, sexuality, 'the good life', care, a haven, unwaged labour, reproduction and immanence.[3] The public *as an ideal type* has traditionally been the domain of the disembodied, the abstract, the cultural, rationality, critical public discourse, citizenship, civil society, justice, the market place, waged labour, production, the polis, the state, action, militarism, heroism and transcendence.[4]

The idea of privacy is deeply embedded in Western political theories of freedom, personal autonomy, patriarchal familial sovereignty and private property. Traditionally there have been spatial and corporeal components to the idea of autonomy. The linkage between individual, family and group autonomy and privatization, localization and other exclusionary spatial strategies is one of the most important and interesting aspects of political geography. However, this linkage is one that is often taken for granted and therefore tends to be naturalized or depoliticized. The idea of spaces (material and metaphorical) hidden from the light of public view in which autonomy is most effectively enacted is widely respected. However, this idea is also highly charged and tension filled for many across the political spectrum.

Lawrence Stone and others have suggested that the perceived need for increased privacy in domestic spaces arose with the European nation-state. Attempts were made by both the state and private households to strengthen the institution of the family and to limit the space of state authority over the reproductive family unit (Stone 1977: 133–42). The home was accordingly considered a microcosm of the political order with the male head of household as ruler.[5] While modern liberal notions of *individual* freedom and rights within the family or household as well as within society clearly differ from these earlier ideas of paternal dominance, the latter are still quite evident in contemporary culture and the administration of justice. As Judith Squires (1994: 394) puts it:

> the preliberal antiliberal patriarchal tradition of family sovereignty has, for reasons not inherent to the liberal tradition itself, been incorporated

– tortuously – into the liberal rhetoric and legislation on privacy rights. Individual autonomy, which is the bedrock of liberal theory, has in practice been conflated with family autonomy.

Historically, in legal terms at least, women have been treated as private and embodied, in the sense of apolitical. They have long been treated as if not fully capable of independent disembodied political thought and objectivity as evidenced by the fact that it was relatively recently that women were given the vote. Still today most men[6] move between public and private spaces and spheres with more legitimacy and physical safety (see Pain 1991; Valentine 1989), and less burdened by responsibilities as caregivers of children and the elderly than most women.

Both private and public spaces are heterogeneous and not all space is clearly private or public. Space is thus subject to various territorializing and deterritorializing processes whereby local control is fixed, claimed, challenged, forfeited and privatized. In some cases this may have socially progressive results in terms of providing a safe base (site of resistance) from which previously disempowered groups may become empowered. On the other hand, isolation in a private or quasi-private space or sphere may have an undesirable depoliticizing effect on a group, fortifying it against challenges from, and allowing it to inadvertantly assume independence from, a wider public sphere. However, as Brian Massumi, in his interpretation of the thought of Deleuze and Guattari, says, there is an important difference between 'entrenching one's self in a closed space (hold the fort)' and 'arraying one's self in an open space (hold the street)' (Massumi 1992: 6). The street serves here as a metaphor for sites of resistance that are part of a rhizome-like process of deterritorializing and a progressive opening up to the political sphere. The fort signifies territories, securely established centres of domination. (On the political ambiguity of place and localizing processes and whether they are conservative or progressive see Massey 1993.)

There are many privatized or quasi-privatized, commercialized public spaces including shopping malls and exclusionary suburbs. This privatization of ostensibly public places has very uneven consequences for the population as a whole because groups with greater resources can more easily privatize spaces. Such privatizing of space is often accompanied by aestheticization as for example when urban space is cleared of marginalized people and political activities and redesigned as a spectacle for the consumption of affluent classes. Furthermore by privatizing (depoliticizing) these spaces, the owners and users of such spaces more easily free themselves from various types of public surveillance, regulation and public contestation.

The private is a sphere where those families who are not dependent on the state for welfare have relative autonomy. Those who *are* dependent, however, are often subject to unwarranted intrusion and surveillance.[7] In general, however, liberal political and legal theory can be seen as a territorializing spatial practice that attempts to differentiate the public and private by erecting a boundary around a private sphere of *relative* non-interference by civil society or the state.

PUBLIC SPHERE

The public sphere is not just the site of state politics and regulation, nor is it limited to the market place or the economy;[8] it is also the site of oppositional social movements. In fact, under many definitions, the public sphere is a political site separate from, and often critical of, the state and the economy.[9] As opposed to the private sphere, it is the discursive and material space where the state and its powers, as well as oppressive aspects of the dominant culture (misogyny, homophobia, racism), are open to challenge by those who have been marginalized in various ways. Don Mitchell gives an example from Berkeley, California.

> The People's Park was working as it should: as a truly political space. It was a political space that encouraged unmediated interaction, a place where the power of the state could be held at bay.
>
> (Mitchell 1995: 110)

According to Mitchell ideally public space is 'unconstrained space within which political movements can organize and expand into wider arenas' (Mitchell 1995: 115). However, he says that most often public space is constituted as 'a controlled and orderly *retreat* where a properly behaved public might experience the spectacle of the city'. In this view public space is seen as politically neutral. Although somewhat more likely to become a site of political organizing than private space, public space is very often planned and controlled for non-political purposes. Public spaces and public spheres often do not map neatly onto one another.

As a normative ideal the public sphere is open to all; in practice, however, it is much more restricted. In fact, Habermas (1991) would argue, the public sphere no longer functions effectively in the interest of any group. Examples of recently increased restrictions on the public sphere as a place where groups can meet to protest and publicize their views is the introduction in Britain of the Criminal Justice and Public Order Act of 1994. This law includes limitations on the right to assemble for peaceful political protests. It is noteworthy that such increased regulation of public behaviour (*allegedly* for fear of *potential* violence) is not matched with a similar increase in the regulation of *actual* violence in the private sphere.[10] This is not to say, of course, that violence in public is adequately controlled or that provisions should not be made to control politically motivated violence.

Moreover, the ideal of a single public sphere that serves as a site of political contestation is considered by some to be either utopian or deceitful in its pretence of homogeneity and inclusiveness (Fraser 1993; Howell 1993; Robbins 1993; Young 1990). There are some very persuasive arguments for the expansion and repoliticization of the notion of public sphere into a multiplicity of heterogeneous publics also known as 'alternative or counter public spheres' or 'counterpublics' or 'critical publics' (Cohen and Arato 1992; Fraser 1993; Robbins 1993). Such counter public spheres can be seen to develop out of social movements. Iris Marion Young (1990: 120) states that:

> the concept of a heterogeneous public implies two political principles:
> (a) no persons, actions, or aspects of a person's life should be forced into

privacy; and (b) no social institutions or practices should be excluded a priori from being a proper subject for public discussion and expression.

Although in practice various critical publics would never be equal in influence or legitimacy, they could ideally all have access to the public sphere and public spaces (where they could challenge, and be exposed to challenges by, members of other counterpublics).

PRIVACY AND DOMESTIC VIOLENCE

Paradoxically the home which is usually thought to be gendered feminine has also traditionally been subject to the patriarchal authority of the husband and father. Personal freedoms of the male head of household often impinge on, or in extreme cases, negate the rights, autonomy and safety of women and children who also occupy these spaces. The designation of the home as private space limits the role of political institutions and social movements in changing power relations within the family. 'A man's home is his castle' – this familiar expression reveals the important historical link between masculinity, patriarchal autonomy and its spatial expression in the form of private property. As a relatively unregulated sphere the private is a place where men have traditionally dominated their families and the privacy to do so has been jealously protected. Legal definitions of privacy thus gender space and tend to reproduce inequalities. As Schneider (1991: 978) put it:

the interrelationship between what is understood and experienced as private and public is particularly complex in the area of gender where the rhetoric of privacy has masked inequality and subordination. The decision about what we protect as private is a political decision that always has important public ramifications.

Although legal ideas of privacy were established to protect civil liberties under certain circumstances they can also:

mask physical abuse and other manifestations of power and inequality within the family . . . The belief is that it is for family members to sort out their personal relationships. What this overlooks is the power inequalities inside the family that are of course affected by structures external to it.

(O'Donovan 1993: 272)

The private home has been historically seen as a place where men have assumed their right to sexual intercourse. Problematic questions of genuine consent on the part of partners have only recently begun to be addressed with any frequency. The private space of the home can also be a place where aggressive forms of misogynous masculinity are often exercised with impunity. It is a place where rape and other forms of non-consensual sexual activity take place more often than many people realize (see Edwards 1989). Although I recognize that the private space of the home is a place where some men use violence as a way to control women,[11] I wish to distance myself from arguments made by radical feminists such as Brownmiller (1975) and MacKinnon (1989) who argue that violence, especially sexual violence, is used by men collectively as a way to control women. This is to implicate many innocent men who abhor violence and it assumes a narrow view of power – one that sees it as primarily coercive. (On this issue see Pain 1991: 425; however, Pain does not take a stand on whether there is a conscious conspiracy among all men and not just sex offenders to intimidate women.)

Instead, I would choose to explore the idea of a complicity which includes men and women who fail to act decisively against both public and private sexual violence, resorting instead to staying at home at night or encouraging wives, daughters and women friends to do so. Here I am not suggesting that individuals place themselves in the 'sucker position' of risking their own safety. Quite the contrary, I am suggesting that such violence directed against women in both public and private spaces is a problem requiring highly organized, structural solutions, not isolated individualistic ones. The feminist slogan 'Take Back the Night' should be seen as a suggestion not for women to disregard personal safety, but for all those who can (not just women) to organize and ask for public funds to transform public spaces to make them safe and accessible to everyone at night as well as during the day.

Because it is very often invisible and inaudible, domestic violence remains a privatized problem. Unanswered questions remain: to what extent is the home an oppressive site of sexual power and pathological types of masculinity? To what extent is domestic violence explained by historically persistent perceptions of masculine autonomy and entitlement within the space of the home? To what extent does the privacy of private property allow or even legitimate misogynistic violence? One reason why the underlying explanations and motivations of domestic violence are unclear is that such abuse has generally been a private and hidden problem. It is a good example of Berger's dictum elaborated by Soja, that it is space more than time that hides the consequences from us (Soja 1989: 22).

Feminists 'discovered' wife beating in the late nineteenth century as they attempted to open up the realm of the private and patriarchal family affairs to public discourse (see Cobbe 1868). Although since the nineteenth century wife beating has been formally outlawed,[12] the issue still does not receive the public attention it deserves. Enforcement of laws is highly inadequate.[13] According to the Surgeon General of the United States, the battering of women by partners and ex-partners is the 'single largest cause of injury to women in the US' accounting for one-fifth of all hospital emergency cases (Zorza 1992: 83). According to the FBI, roughly 6 million women are abused and 4,000 women are killed by their partners or ex-partners in America each year (Saland 1994). These statistics suggest that domestic violence cannot be dismissed as something private and beyond the scope of public responsibility (Thomas and Beasley 1993: 45). Clearly there are contradictions between ideas of privacy, which assume autonomy of male heads of household, and the prevention of 'the violence of privacy' (Schneider 1994).

Police officers in many places are given a great deal of discretion in dealing with 'domestic disturbances'. Often such 'domestic' calls are not taken seriously. When the police do go to a house, they usually do not make an arrest. Wives may not decide to press charges fearing the alien world of courts, police stations and publicity, even more than the familiar, private violence of the home. The police may understandably fear for their own safety or even their lives.[14] Furthermore the police sometimes share the misogynous views of the batterer, believing, for example, in corporal punishment for 'nagging' wives or at least sympathizing with an overstressed husband.

Many programmes to aid battered women have focused on establishing outside moral as well as material support in order to counter the batterer's often strenuous attempts to privatize the problem by cutting his partner off from contact with relatives, friends and public institutions (Pence and Shepard 1988: 291). One can clearly see spatial strategies at work in the abuser's attempt to isolate his partner from extended family and other social networks by confining her in private spaces. This makes it difficult for her to seek outside support or to organize politically with other battered women. Answers may lie, in part at least, in deterritorializing public and private spheres – that is in questioning the links between individualism, privacy, autonomy and allegedly apolitical private spaces.

Making contacts and establishing outside support networks is a crucial step for a woman who seeks to escape a violent home. It is often difficult to break the financial and emotional dependence on the family home and husband. The need for alternative housing is paramount. Women's shelters often provide temporary accommodation. Daycare, peer and professional counselling, and various training programmes are sometimes available through such organizations.

Women's shelters provide a site of resistance against the imprisoning strategies of the battering partner. While the names of shelters sometimes convey the idea of much-needed social networks – Friends of the Family, Woman to Woman and Good Neighbors Unlimited – often they reflect this spatial dimension – Womanspace, Women's Survival Space, Safe House and Safespace.

Beyond the lack of sufficient funding for shelters and limited space availability, and beyond psychological and economic dependence of women on abusive partners, there are many other reasons why shelters do not always provide an effective solution to isolation and violence. Many woman do not take advantage of the opportunity to remove themselves physically from violent situations for various geographical reasons. Women from rural areas may have to travel long distances to find a shelter and women from non-English-speaking communities may be reluctant to leave a neighbourhood where they have some degree of language and cultural support. However, a much more pervasive sentiment that affects not only the willingness of a woman to go to a shelter, but also interest in funding such shelters, is the individualism and privatism of British and North American society. People who must depend on the help of strangers often feel shame. The ideal of the private family home is so deeply ingrained that even temporary residence outside of such private spaces can be highly embarrassing and stigmatizing. The idea of communal living and sharing of tasks which is encouraged in such shelters is unfamiliar. The fact that shelters are outside the norms of Anglo-American society is also reflected in the names of the shelters – many of which make reference to the temporary, crisis-induced nature of these shelters: Women in Crisis, Transitional Living Center, Assault Crisis Center, Crisis Intervention, Guest House, Emergency House, Sojourn Women's Center, Victim's Crisis Center and Women's Transition House.[15]

While many feminists wish to expose the abuses of masculine privilege in the home, others worry about the opening up of the private to public surveillance, because it could simultaneously open up the realm of individual reproductive rights to state interference. However, Elizabeth Schneider argues for a right to privacy that is not 'synonymous with the right to state non-interference with actions within the family' (1994: 53). Recalling Justice Douglas' opinion in *Roe v. Wade* (1973) she suggests that the concept of privacy has the potential to be defined affirmatively as the right to autonomy for all family members which requires freedom from battering and coercion by partners.

Judith Squires also argues for a notion of privacy closely linked to individual autonomy on empowerment. She writes:

> there are very strong grounds for articulating a specifically embedded and embodied conception of privacy as a means of conferring autonomy. For the body can be viewed as one of the core territories of the self: control over one's own body is crucial to the maintenance of a sense of self and hence the ability to interact openly with others. To have control over own's bodily integrity (to regulate access to it) and to have this integrity recognised, is a minimal precondition for free and equal social interaction. To ensure the possibility of such an embodied autonomy for all persons in contemporary society – with all its multifarious mechanisms of observation and control – we will need a political defence of privacy rights.
>
> (1994: 399)

Others argue for the necessity of private spaces for protection against an overly aggressive state. However, despite this very real consideration, I would argue that the existence of relatively unregulated spaces is a political arrangement that tends to hide the causes as well as the consequences of oppressive power relations within the family from a wider public. It protects particularly those who have the resources to most effectively privatize space. Intrusive or even fascistic state practices (and all their hidden and privatized manifestations) might better be opened up to scrutiny in the public sphere (or counter public spheres) under more informal, unconstrained and inclusive conditions of discourse and debate within civil society.

I suggest, then, that there is a positive concept of privacy related to the autonomy of individuals which allows for and may even require the opening up of private spaces to the public sphere in order to protect individuals whose autonomy is compromised by the concept of unregulated private space, especially when that space is constituted by unequal power relations or outmoded ideas of domestic patriarchal sovereignty.

Early feminism concentrated so much energy on opening up the public sphere to women through the use of sex-discrimination legislation that the question of how the private sphere might be reconstituted through law was rarely addressed. Only a few basic steps have been taken in this direction. An example is the elimination of spousal exclusion from the possibility of rape.

A broadly Foucauldian conception of (albeit highly uneven) relations of power as suffused throughout society and across space can aid in undermining the public/private dichotomy. 'The personal is the political' is a proclamation commonly heard among feminists, gays and lesbians that challenges the public/private dichotomy as it has traditionally been formulated. This phrase serves as an evocative reminder of the artificiality of such a clear-cut distinction despite its long history and naturalization in legal discourse. It is a statement of the fact that personal relationships are also power relationships and that everyone is implicated in the production and reproduction of power relations. Domestic and even intimate relations are political relations that can be transformed through political means. Although places may be more or less overtly politicized, there are no politically neutral spaces. Similarly, whether or not embodiment is explicitly recognized – whether or not a *dis*-embodied, allegedly objective perspective is claimed – the spatial and social situatedness which comes from necessary corporeality is inescapable. Foucault (1980: 187) argues power relations emanate not only from state or juridical sources, but concern:

> our bodies, our lives, our day-to-day existences . . . Between every point of a social body, between the members of a family, between a master and his pupil . . . there exist relations of power.

Furthermore, Foucault argues, such power is met with a multitude of points of resistance throughout a network that encompasses the whole of a society.

One important form of resistance is to bring issues of privatized power relations into a public forum where efforts to bring about structural change in these relations can be more easily organized. There has always been a close relation between the degree of women's confinement in private space and their relative exclusion from the public sphere of organized social movements and political action. Thus a smoothing out of the public/private boundary and an opening up of privatized problems to public contestation is necessary despite risks of facilitating undue state intervention. Non-progressive state intervention into private lives can theoretically be prevented through the increased use of the critical functions of publicity and strengthened, increasingly heterogeneous public spheres.

PRIVACY AND RESISTANCE

Doreen Massey (this volume) speaks of the problematic boundary between workplace and home as reinforcing the gendered distinction between transcendence and immanence. Transcendence is the use of Reason in the production of History, Knowledge, Science and Progress; immanence is 'the static realm of living-in-the-present', of reproduction, of servicing those who make history. Contrasting this public/transcendence private/immanence correlation in the US and Britain with examples from Eastern Europe, Joanne Sharp (this volume) points to cases in which the identification of private space with immanence and public space with transcendence was inverted. During the communist period a major site of resistance and political organization was in the private space of the home. Civil society was spatially marginalized by powerful governments forcing it into a repoliticized private sphere of the home: 'It was here rather than in any formally public sphere that the possibility for transcendence occurred.' She adds, however, that the opposition of family to state served only to deflect attention away from uneven power relations within the family.

bell hooks offers another destabilizing perspective on the idea of the traditional home as a place of immanence rather than transcendence. She says that because public space can be very hostile to African Americans (men as well as women), the home can be an important site of resistance. She sees the homeplace as having a radical political dimension. It's a place where, as she says, 'we could restore to ourselves the dignity denied us on the outside in the public world' (hooks 1990: 42). hooks, however, makes it clear that while the home *can* be a site for organizing subversive activity, it is often viewed as a politically neutral space where the political role of black women is devalued (hooks 1990: 47). She blames the influence of white, bourgeois norms (which produce domestic space as an aestheticized space of consumption and reproduction) for redefining the home as a depoliticized site.

hooks acknowledges of course that the black home can also be the site of patriarchal domination. Kimberlé Crenshaw sees it as a site of multiple oppressions where women of colour sometimes face an 'intersectional dis-empowerment of race and gender'. She states that women of colour who are subjected to domestic violence are often reluctant to call the police as there is:

> a general unwillingness among people of color to subject their private lives to the scrutiny and control of a police force that is frequently hostile. There is a more generalized community ethic against public intervention, the product of a desire to create a private world free from the diverse assaults on the public lives of racially subordinated people.
>
> (1994: 103)

But, as Crenshaw states: 'this sense of isolation compounds efforts to politicize gender violence within communities of color, and permits the deadly silence surrounding these issues to continue' (1994: 111).

Nevertheless, hooks argues for the need to reaffirm the home as a site of organizing, affirming political solidarity and regrouping for resistance in spite of the fact that from the standpoint of the relatively more powerful this may seem a minor political resource. hooks (1990: 45) argues that: 'the devaluation of the role black women have played in constructing for us home-places that are the site for resistance undermines our efforts to resist racism and the colonizing mentality which promotes internalized self-hatred'.

Habermas (1991) also sees certain private sphere institutions as having served, in the past at least, important political purposes. He points to the literary salon, club, cafe and lodge as semi-private political spaces with a public sphere critical function. I cite these various examples to show that there is often no clear-cut distinction between the private as a site of immanence and the public as a site of transcendence. These examples should not be interpreted as showing that private spaces (which have a tendency to be exclusionary and isolating) are ideal sites of liberation struggle, however.

Supportive home environments can, of course, also reproduce white racism. Iris Marion Young argues that private, homogeneous, and exclusionary spaces provide autonomy that should be distinguished from empowerment. While she sees autonomy as a closed concept referring to non-interference, empowerment is an open concept allowing agents to participate in democratic decision making (Young 1990: 251). Possibly a distinction should be made between private spaces that are sites of empowerment and resistance (becoming open, publicized and political) and private territories that are exclusionary or oppressive (remaining closed and private in the sense of spaces where the privacy of some to oppress others – who for various reasons may share the privatized spaces – is protected from public or state regulation). This distinction may be useful in conceptually opening up the boundary between the public and private.

THE SPATIAL REGULATION OF HOMOSEXUALITY

Like gender, sexuality is often regulated by the binary distinction between public and private. It is usually assumed that sexuality is (and should be) confined to private spaces. This is based on the naturalization of heterosexual norms. Naturalized heterosexuality makes sexuality in public spaces nearly invisible to the straight population (Valentine 1993). Surveys have shown that the majority of respondents have no objection to homosexuals as long as they 'do not flaunt their sexuality in public' (Herek 1987 as quoted in Valentine 1993). 'What they do in private is nobody's business', is a commonly heard, well-intentioned expression. However, as Gill Valentine puts it, the idea of homosexuality as appropriate only to private spaces:

> is based on the *false* premise that heterosexuality is also defined by private sexual acts and is not expressed in the public arena . . . This therefore highlights the error of drawing a simple polar distinction between public and private activities, for heterosexuality is clearly the dominant sexuality in most everyday environments, not just private spaces, with all interactions taking place between sexed actors.
>
> (1992: 396)

While public space appears heterosexist to gays and lesbians, many expressions of sexuality are so naturalized as to be virtually invisible to the straight population. As Valentine points out: 'heterosexuality is institutionalized in marriage and the law, tax, and welfare systems, and is celebrated in public rituals such as weddings' (1992: 396).

Valentine and others have pointed out that suburban housing developments as sites of overtly heterosexual as well as familial sentiments and rituals are generally considered alienating environments by lesbians and gays. While the home may be a haven for some gay couples, the family home is often an extremely heterosexist and alienating site for gays.

Gloria Anzaldua speaks of a former student, a lesbian, who said that she connected the word homophobia with 'fear of going home' (Anzaldua 1987: 20). Here, in the home, the patriarchal, heterosexist exercise of territorialized power and regulatory practices freed from public intervention and political contestation may be especially threatening, keeping gay identities in the closet. The spatial metaphor of the closet is a particularly telling one in this context where gays may not be 'out' even to their own families within their own home.

Although many would think of workplaces as generally asexual (except for occasional sexual harassment) these are nevertheless also heterosexual and often heterosexist spaces. Nearly invisible because it is universalized and naturalized, heterosexuality is inscribed in public as well as private spaces as the dominant ideology. Like trying to convince WASPs (White Anglo-Saxon Protestants) that they have an ethnicity, it is difficult to make heterosexuals aware that their spaces invoke a sexuality. Naturalizing one's own hetero-sexuality means imposing one's own inability to see him or herself as Other on one's surroundings. Failing to notice your own difference as heterosexual is an act with significance. It leads to the heterosexing of space.

An interesting article by Bell *et al.* (1994) addresses the issue of various ways the heterosexuality of public space might be resisted. The authors examine the performance of two types of homosexual identities, the hyper-feminine 'lipstick lesbian' and the hypermasculine gay 'skinhead' that serve (often unintentionally) to parody heterosexual identities. They ask, however, whether such stylistic transgressions of popularly held stereotypes of lesbians and gays can actually have any significant destabilizing effect on heterosexism and the assumption of public spaces as generally asexual. They worry about the danger of celebrating transgression for transgression's sake. I share the latter concern and argue that there is a danger of the aestheticization of politics whenever style is used as a mode of transgression. However, I argue, once again, that significant social change requires organized action in the public sphere and access to various resources, including the media, rather than individualistic, privatized action.

Thus, I suggest that lesbian and gay practices which potentially denaturalize the sexuality of public places could be more effective if they were widely publicized. If they were made more explicit and readable then contests around sexuality would become more visible to the straight population. Furthermore, one would expect that such denaturalizing tactics would work for the gay population as well, by pointing to the fluidity of identity and helping to transgress clear-cut heterosexual/homosexual dichotomies including stifling codes of dress and behaviour sometimes imposed in an attempt to stabilize an internally coherent identity politics.[16] The media could also do more to publicize some of the complex and challenging questions about the performance and the reconstitution of gender and sexuality in public spaces that are raised in this article and in the work of Judith Butler (1990, 1994) upon which Bell *et al.* (1994) draw.

Public space can be used as a site for the destabilization of unarticulated norms, or as Munt calls it, the 'politics of dislocation' (Munt 1995: 124). Deconstructive spatial tactics can take the form of marches, Gay Pride parades, public protests, performance art and street theatre as well as overtly homosexual behaviour such as kissing in public. An example cited by Bell and Valentine (1995) of such tactics is the 'queering' of space by Queer Nation Rose (QNR) and ACT-UP who refused to allow the Montreal Pride Parade to be ghettoized in the gay village as it had in past years. Instead they marched through the downtown streets. Furthermore, they ignored the anti-drag, anti-leather parade rules by declaring 'If you're in clothes you're in drag': 'irreverent combinations of identities proliferated, including fags posing as dykes, dykes dressed as clone fags, and bisexuals pretending to be fags pretending to be lipstick lesbians' (Bell and Valentine 1995: 14).

Tim Cresswell makes the important point that it is difficult to get people to recognize normative geographies until these are transgressed. 'By looking at events which upset the balance of common sense', he says:

I let events themselves become the questions. The occurrence of 'out-of-place' phenomena leads people to question behavior and define what is and what is not appropriate for a setting. The examination of commonsense becomes a public issue in the speeches of politicians and the words of the media.

(Cresswell, 1996)

When spatial tactics of queer politics become what Cresswell (in another context) calls 'crisis points in the normal functioning of everyday expectations' for the mainstream heterosexual population – then normative heterosexual geographies become more clear. This is the first step towards destabilizing and eventually overturning such repressively striated geographies of gender and sexuality.

SPATIAL MARGINALIZATION OF SEX WORKERS

Prostitutes also offend the aesthetic sensibilities of the upholders of the public/private dichotomy. They upset the 'everything in its place' mentality that reproduces the public/private spatial dichotomy. They threaten notions of 'respectable' and 'orderly' behaviour on the part of women who, it is thought, should be escorted at night in public spaces. Because of women's traditional exclusion from the political sphere, the term 'public woman' in dominant discourse has traditionally meant 'not respectable', a prostitute, whereas a public man was a statesman (Matthews 1992). To be a respectable woman was to sexually serve one man – a husband at home. While this ideal need no longer be strictly adhered to for a woman to be considered respectable in Anglo-American society, all forms of commercial sex are generally considered beyond the bounds of respectability.

Glenna Matthews (1992) surveys many sites of resistance against this definition of public woman that have accompanied the rise of women in the public sphere. However, there are other agents and sites of resistance against the gendered public/private dichotomy. There are many women with a strong sense of agency who are proud to be public women in the traditional sense of the term. In many cases they would not wish to join the ranks of establishment feminism or the political elite. Many (but certainly not all)[17] adult prostitutes and other sex workers freely choose[18] marginal or eccentric locations from which to claim their rights as sexual minorities and challenge the very structures which elite women employ to get ahead. They also challenge the narrow definitions of politics and power employed by those who seek public office. However, their views have only just begun to be heard as they have long been silenced by members of the dominant culture, including many prominent feminists who see their work and lifestyles as epitomizing oppression by men.

Prostitution is a good example of a practice both spatially and socially marginalized by societal attitudes and the law. There are complex spatial implications in the laws regarding its practice. The laws in Canada and Britain make prostitution itself legal in principle but all but impossible to practice without breaking one of many laws. These laws regard such issues as solicitation or procuring in public places, where prostitution may be practised and who may benefit from the profits gained.[19] This latter restriction makes it illegal for a prostitute to live with members of her family if they benefit from her earnings.

These externally imposed spatial limits to the legal practice of prostitution again deny the sexuality of public places by imposing greater spatial restrictions on sexual minorities than on those who conform to the societal standards. In some places these limits serve to hide from public view and thus privatize many of the aesthetically and morally offensive physical, psychological, medical and social problems surrounding the highly marginalized identities of prostitutes. In other places, they force prostitutes onto the street where they can be subjected to surveillance and segregating practices of the police.

The state and public morality (the latter represented by the religious right and certain radical feminists[20] among others) define prostitutes as either deviant and immoral or victims suffering from false consciousness who symbolize the oppression of all women by all men. Such characterizations succeed in cutting them off from having an effective role as public women in the political sense of speaking on their own behalf and reclaiming their civil rights. These include their right to citizenship,[21] to work in safe conditions, their right to exercise control over their own bodies and to earn respect as healers, sex experts and business women as well as the right to freedom from harassment by police and self-proclaimed upholders of public morality (Bell 1994: 100).

If prostitutes could safely 'come out' in the public sphere and speak on their own behalf there would be many benefits, including the opportunity to add some new, knowledgeable voices to the debates over the meaning of choice and consent, personal autonomy, sexual exploitation, victim identities, false consciousness and power relations, structural explanations for what are all too often seen as individual problems.

CONCLUSION

To conclude, I would say that the public/private distinction is still among the most important spatial ordering principles in North America and Britain today. Public space is regulated by keeping it relatively free of passion or expressions of sexuality that are not naturalized, normalized or condoned. It is further regulated by banishing from sight behaviours that are in various cases repugnant either rightly (as in the case of domestic violence) or wrongly (as in the case of publicly expressed homosexuality or unforced adult prostitution) to many members of the dominant groups in society. The institutionalized dividing off of critical public debate and political expression into specialized and increasingly controlled spaces allegedly allows for the possibility of disembodied dispassionate rational discourse and formal political decision making under conditions of public order. This has left the private domestic sphere to remain invisible, relatively unregulated (i.e. selectively regulated) and generally free from public scrutiny. However, we argue that certain so-called private issues need to be deterritorialized, that is more thoroughly public(ized) and legitimated as appropriate to public discourse. As Benhabib puts it, 'the struggle to make something public is a struggle for justice' (1992: 94). This should *certainly* not be taken to mean that justice is *necessarily* served when an issue becomes publicized. It is, however, more *likely* to be served in a truly open public debate where no parties affected by an issue are excluded.

Subversive discourses first articulated in private spaces may eventually become public. Members of various marginalized social movements eventually learn to negotiate their way into the public sphere. However, feminist political practice has begun to tackle the problem of the public/private sphere distinction *itself* as a gender-biased spatial practice which facilitates what are largely gender-specific abuses and also the marginalization and enforced privatization of sexualities which do not conform to dominant ideas of 'natural' dynamics of heterosexual love. Their explicit intent is to reveal exactly how disempowering it can be for those who differ from the *allegedly* neutral norms and therefore cannot act with the same degree of autonomy or protection assumed by established models of the democratic society.

The goal is to mount a multi-pronged attack on the spatial and discursive boundaries that regulate behaviour and discipline difference. This would entail among other things an 'outing of everybody'. By 'outing' here I do not refer to the highly problematic practice of the outing of individual gays. I think that the practice of publicly identifying the sexual orientation of individuals against their wishes cannot be considered a just or effective solution to the problem of homophobia. Nor do I mean to say that privacy should not be respected when it does not harm others. Rather, I suggest that the boundaries between the private and public can be destabilized by being actively questioned and placed in the public consciousness through the media, through challenges in the courts and through the efforts of social movements. The physical design of our societies' highly privatized landscapes however, have been shaped not only to protect those whose privacy should rightfully be respected, but also to secure the privacy and autonomy of the abusers of the women and children who share their domestic spaces.

By 'outing', then, I am talking in very general terms about a transformed spatiality – an empowering deterritorialization, the creation of smooth, less striated space. Here I refer to Deleuze and Guattari's (1987) notions of smooth or open-ended space as opposed to state space which they describe as striated or gridded. Smooth space is contrasted with defended, exclusionary, confining spaces where oppressive patriarchal and heterosexist practices can become entrenched. These terms are highly abstract and meant to be evocative. Allowed to wash over one, listened to like music, as Deleuze and Guattari suggest, their writings provide inspiration to rethink conventional notions of space. I refer to a potential spatial revolution that would conceive of physical and political or discursive space as less clearly divided between publicly recognized territories of formal power, depoliticized spaces of urban spectacle and protected domestic spaces of uneven privatized power relations. This would enable the consequences of our individual and collective actions to be made more visible and accountable to critical public debate and oppositional social movements. To quote Seyla Benhabib once again:

> All struggles against oppression in the modern world begin by redefining what had previously been considered 'private,' non-public and non-political issues as matters of public concern, as issues of justice, [and] as sites of power.
>
> (1992: 100)

Although privacy has always been a contingent rather than an absolute right, it is widely cherished and seen as indispensable for the protection of individual autonomy. However, privacy is closely associated with highly privatized spatial arrangements and social codes of 'civil inattention' which facilitate the violation of the rights of a significant percentage of the population. Thus we must stop to ask ourselves if there are not better ways to control the abuses of state and other public manifestations of power. Should each individual and social movement be left to individually renegotiate the public/private spatial and discursive boundaries for themselves? Or should this deeply rooted division (so sacred and central to understandings of personal freedom) be radically rethought? I would argue for the latter.

I do not endorse spatial or political anarchism. There is clearly a need for effective government at a range of scales (see Penrose 1993: 46), various types of regulation, a progressive redistribution of power and resources, and expanded, multi-scale welfare programmes. However, the ideal geography would work to minimize: household autonomy as opposed to the empowerment of its individual members, place-based identity and privilege, local control which has highly uneven consequences for social justice across communities,[22] nationalism, and other territorializing and confining exclusionary processes. The creation of progressive geographies would require deterritorialization – the creation of open-ended, proliferating and inclusive sites of empowerment and resistance against exclusionary, reterritorializing processes: place essentialism and homogenizing identity politics or coerced assimilation. These would be sites of 'radical openness' as bell hooks (1990) puts it – sites which may be nurturing – which may serve as havens, but which are opened up to the public sphere and politicized (or repoliticized) as the case may be. On the other hand while deterritorialized geographies would encourage heterogeneity they would also discourage the naturalization, reification and ghettoization of differences – including, importantly, differences of gender and sexuality. Fluid geographies would construct and in turn be constructed by fluid identities.

ACKNOWLEDGEMENTS

An earlier version of this paper was the introduction to a three-part presentation by Lynn Levey, Beth Wolgemuth and myself for the symposium 'Place, Space and Gender' held at Syracuse University in the Spring of 1994. I would like to thank Beth for ideas and inspiration especially on the sex worker section and Lynn whose unpublished paper 'No place like home: shifting legal paradigms to make women safer' provided me with references and information on domestic violence. I would also like to thank Joanne Sharp, Michael Landzelius and two anonymous reviewers for commenting on an earlier draft.

NOTES

1 A study which looks at the homeless as an example of another marginalized group which has developed spatial strategies that transgress the public/private distinction, politicize space, and attempt to claim sites of resistance against the regulation of behaviour in public spaces is Mitchell (1995).
2 Gayatri Chakravorty Spivak (1988: 103) consequently goes so far as to claim that 'the deconstruction of the opposition between the private and the public is implicit in all, and explicit in some, feminist activity'.
3 On immanence as distinguished from transcendence see de Beauvoir (1974), Lloyd (1984) and Massey (this volume).
4 There are in fact a number of different public/private distinctions; these include the state versus the market, citizenship versus both the state and the economy, and domestic versus waged labour. See Robbins (1993: xiii) who draws on an essay by Jeff Weintraub.
5 It should be noted that over the centuries the role of the male head of household as well as the very notion of masculinity itself has varied considerably, and it differs by class as well. See Tosh (1994) for a review of the literature on this subject.
6 But see Myslik (this volume) on the limitations placed upon the free movement of gay men by those who harass and violently attack them.
7 Such intrusion points to the downside of increased state regulation of the private sphere. One would hope however that this might be rectified through more enlightened public policies which distinguish between areas of public concern and people's legitimately personal affairs.
8 As stated in note 3, the public is sometimes defined as the state in opposition to civil society, sometimes the public includes the market. The market sometimes is seen as private, however.
9 On the history – and normative theory – of the public sphere see Habermas (1991). For debates around Habermas's concept see Calhoun (1993) and Robbins (1993).
10 Ironically, while rape and assault is often ignored if it takes place in private spaces,

consensual and private sexual practices among gays are sometimes not tolerated. Operation Spanner was a recent police operation in Britain designed to catch and arrest men participating in sadomasochistic activities in private spaces. Sixteen arrests were upheld in court on the grounds that sadomasochistic practices among consenting adults can not be afforded protection by laws of privacy (on this see Bell 1995: 305).

11 In fact it was not until the twentieth century (1922) that wife beating had become illegal in all US states (Pleck 1987: 108–21).

12 For example in Britain the Matrimonial Clause Act which dates from 1878 (see Hammerton 1992).

13 Buel (1988: 217) states that police officers fail to arrest in the majority of cases where battered women request an arrest be made. Some policemen say that they arrest depending upon the reason the man hit his partner, perpetuating the notion that some women deserve to be beaten.

14 According to the United States Commission on Civil Rights (1982: 12–22) a majority of the police who have been killed on duty were handling 'domestic disturbance' cases.

15 Women's shelters first opened in Britain in 1971, spread to Europe and then to the US. It is not surprising that the US would be slower to accept the idea of shelters in that individualism and privatism are even stronger than in Britain. And the public spheres are weak in comparison with Britain.

16 Oppressive dress codes was the topic of a paper by Gill Valentine presented at the Association of American Geographer's national meeting in San Francisco, April 1994.

17 The World Charter of Prostitutes' Rights distinguishes between voluntary and coerced prostitution: 'Voluntary prostitution is the mutually voluntary exchange of services for money or other consideration; it is a form of work, and like most work in our capitalist society, it is often alienated, that is, the worker/prostitute has too little control over her/his working conditions and the way the work is organized. Forced prostitution is a form of aggravated sexual assault' (quoted in Bell 1994: 114). It calls for the decriminalization of all aspects of voluntary adult prostitution. The Charter also states the need for help and retraining for prostitutes wishing to leave prostitution. It states, 'The right not to be a prostitute is as important as the right to be one' (quoted in Bell 1994: 116).

18 By using the words 'freely choose' here I am not suggesting any kind of radical freedom. Freedom to choose work in a capitalist society is of course highly contingent. Most choices are accompanied by some degree of alienation and contradictory consciousness. Furthermore, prostitutes typically (but not always) have fewer choices than the majority of individuals in society.

19 Prostitution is illegal in 49 out of 50 states of the US.

20 The MacKinnon–Dworkin wing of feminism is often referred to as radical feminism. It is known for its campaigns against pornography and prostitution and their affinity with the organization WHISPER (Women Hurt in Systems of Prostitution and Engaged in Revolt). Such feminists construct prostitutes as victims of male oppression by definition and thus seek to end prostitution. They stand in opposition to prostitutes' rights groups which seek to empower prostitutes and politicize sex work (Bell 1994: 99–102). The latter are represented by the International Committee for Prostitutes' Rights and two World Whores' Congresses and groups such as the San Francisco-based COYOTE (Call Off Your Old Tired Ethics) and Toronto-based CORP (Canadian Organization of Prostitutes' Rights).

21 Various prostitutes' rights groups in North America and Europe have been campaigning for legal and human rights including freedom of speech, travel, immigration, work, unionizing, marriage and motherhood, employment insurance, health insurance and housing. *The World Charter for Prostitutes' Rights* seeks decriminalization of 'all aspects of adult prostitution resulting from individual decision' (i.e. based on either free choice or necessity) (see Bell 1994: 113).

22 This is especially true in the US as opposed to Britain. In Britain local government is far less dependent upon locally generated funding for various community projects; local control means control over funds largely generated at the national as opposed to the local level, thus there is not the same inequality between communities.

(RE)NEGOTIATING THE 'HETEROSEXUAL STREET'

Lesbian productions of space

Gill Valentine

THE HETEROSEXUAL STREET

In November 1991 a lesbian couple made the headlines in the British gay press when they were thrown out of a supermarket in Nottingham for kissing in the store (*Scene Out* 1991). What their experience demonstrates is that the street[1] – and I mean this to include not only the pavement/sidewalk, but also the places, such as shops and cafes, which the street contains – is not an asexual space. Rather, it is commonly assumed to be 'naturally' or 'authentically' heterosexual (Bell *et al.* 1994). Whilst couples of the opposite sex are free to embrace over the supermarket trolley, the lesbian kiss caused panic because 'images of selves trouble as they cut into spaces where they don't "belong"' (Probyn 1992: 505).

Judith Butler has famously argued in her book *Gender Trouble: Feminism and the Subversion of Identity* that: 'gender is the repeated stylization of the body, a set of repeated acts within a highly rigid regulatory framework that congeal over time to produce the appearance of substance, of a natural sort of being' (1990: 33). In the same way the heterosexing of space is a performative act naturalized through repetition and regulation (Bell *et al.* 1994; Bell and Valentine 1995). This repetition takes the form of many acts: from heterosexual couples kissing and holding hands as they make their way down the street, to advertisements and window displays which present images of contented 'nuclear' families; and from heterosexualized conversations that permeate queues at bus stops and banks, to the piped music articulating heterosexual desires that fill shops, bars and restaurants (Valentine 1993). These acts produce 'a host of assumptions embedded in the practices of public life about what constitutes proper behaviour' (Weeks 1992: 5) and which congeal over time to give the appearance of a 'proper' or 'normal' production of space.

Whilst heterosexuals have the freedom to perform their heterosexuality in the street – because the street is presumed to be a heterosexual space – sexual dissidents, as the Nottingham lesbians found out, are only allowed 'to be gay in specific spaces and places' (Bristow 1989: 74). Whilst the space of the centre – the street – is produced as heterosexual, the production of 'authentic' lesbian and gay space is relegated to the margins of the 'ghetto' and the back street bar and preferably, the closeted or private space of the 'home' (although even this is not always acceptable, as David Bell, 1995, argues concerning the complex ways that the state regulates sexual citizenship). Thus the London Lesbian Offensive Group claim that 'Heterosexual privilege is about having, and assuming, the right to be more "normal" in both public and private. (Public not meaning outside your home, but in absolutely all dealings with the everyday world)' (London Lesbian Offensive Group 1984: 257).

But the production of heterosexual space is not only tied up with the performance of heterosexual desire but also with the performance of gender identities. Despite Gayle Rubin's claim that 'it is essential to separate gender and sexuality analytically to more accurately reflect their separate social existence' (1984: 308), it is hard to escape the role gender identities play in the production of heterosexual space. On the one hand, gender and sexuality are *not* the same thing but on the other hand they are certainly closely related, mutually constitutive perhaps, as Vron Ware (1992) has argued about gender and race. Certainly, Judith Butler makes a convincing case for the argument that binary gender identities only make sense within a heterosexual framework or matrix. She writes:

> The institution of a compulsory and naturalized heterosexuality requires and regulates gender as a binary relation in which the masculine term is differentiated from a feminine term, and this differentiation is accomplished through the practices of heterosexual desire. The act of differentiating the two oppositional moments of the binary results in a consolidation of each term, the respective internal coherence of sex, gender and desire.
>
> (Butler 1990: 22–3)

The specific 'feminine' 'shape' and 'look' that is perceived as heterosexually desirable and that is (re)presented and (re)produced through the bio-power channels of fashion, health, diet, fitness and so on (Evans 1993) may change over space and time but essentially being a woman is about performing a gender identity that is perceived to maintain the unity or coherence of gender, sex, desire by articulating a discrete asymmetrical opposition between the 'feminine' self and 'masculine' other (so that, to paraphrase Rosalind Coward (1984: 42) 'sexuality pervades our bodies almost *in spite of ourselves*').

Lesbians and gay men have historically been assumed to have 'twisted' gender identities, so that gay men are labelled as effeminate and lesbians as butch just as effeminate men and masculine women are perceived to be gay. This despite plenty of evidence to the contrary; for example, lipstick lesbians are the embodiment of femininity and 'many heterosexuals are not respectively masculine and feminine, or not in certain respects all the time' (Sinfield 1993: 22). Thus repetitive performances of hegemonic asymmetrical gender identities, like repetitive performances of heterosexualities, also produce a host of assumptions about what constitutes 'proper' behaviour/dress in everyday spaces which congeal over time to produce the appearance of 'proper', i.e. heterosexual, space. In this way 'the lesbian subject is always a doubled subject caught up in the doubling of being a woman and a lesbian' (Probyn

1995 : 81). For example Linda McDowell's (1995) study of merchant bankers demonstrates the way that working in the city involves the construction of an embodied gender performance, in which attributes of masculinity and femininity, including a more or less authentic presentation of sexual identity, are not only an integral part of doing business but also produce the bank as a heterosexual space.

But sex, gender and desire do not necessarily map coherently onto each other to maintain the logic of heterosexuality. As Alan Sinfield argues 'ideological categories fail to contain the confusions that they must release in the attempt to achieve control. That is why we observe heterosexuality plunged into inconsistency and anxiety; it is aggressive because it is insecure' (Sinfield 1993: 22). This insecurity often manifests itself in the form of regulatory regimes that constrain the possible performances of gender and sexual identities, in order to maintain the 'naturalness' of heterosexuality. These are regimes which take the form of multiple 'processes, of different origin and scattered location, regulating the most intimate and minute elements of the construction of space, time, desire and embodiment' (Foucault 1979: 138).

One such process, as the Nottingham lesbians trying to shop for groceries found out, is the simple removal of those who cut into and disrupt the 'normality' of heterosexual space by performing their desires in a way that produces (an)other space. In the UK, for example, a number of statutory and common laws can be and often are, used to criminalize public displays of same sex desire on the streets. Although these laws do not explicitly single out lesbians and gays, they are often interpreted and applied in an extremely discriminatory way against sexual dissidents (Foley 1994). Public order laws, for example, have been invoked to threaten a couple walking hand in hand with prosecution, and have been used to obtain a conviction for 'insulting behaviour' against two sexual dissidents seen kissing at a bus stop in the early hours of the morning (Foley 1994).

Often, however, anxious straight citizens don't wait for the police or private security forces to step in and stabilize the heterosexuality of the street, rather they actively regulate it through aggression (Herek 1988). As Virginia Apuzzo, former executive director of the National Gay and Lesbian Task Force has pointed out, 'To be gay or lesbian in America is to live in the shadow of violence' (Comstock 1991: 54). For example, levels of victimization reported by lesbians in a survey in Philadelphia were twice as high as those recorded for women in the general urban population (Aurand et al. 1985; Comstock 1989, 1991). Dr Stewart Flemming describes the type of injuries sustained by sexual dissidents who are brought into his San Francisco medical centre for treatment as:

> vicious in scope and the intent is to kill and maim . . . Weapons include knives, guns, brass knuckles, tire irons, baseball bats, broken bottles, metal chains, and metal pipes. Injuries include severe lacerations requiring extensive plastic surgery; head injuries, at times requiring surgery; puncture wounds of the chest, requiring insertion of chest tubes; removal of the spleen for traumatic rupture; multiple fractures of the extremities, jaws, ribs and facial bones; severe eye injuries, in two cases resulting in

permanent loss of vision; as well as severe psychological trauma the level of which would be difficult to measure.

(Comstock 1991: 46)

Whilst gay men are primarily attacked by other men, the perpetrators of violence against lesbians include not only men (alone and in groups with other men and/or women), but also women (alone and in groups). There are also gender differences in the geography of homophobic assaults – lesbians report more violent encounters in the 'heterosexual street' than gay men, who are more likely to be victimized in cruising areas, or gay-identified neighbourhoods (Berrill 1992; Comstock 1991; Valentine 1993).

In many cases these incidents are not 'provoked' because lesbians are articulating their sexuality, for example by kissing or cuddling, but rather because they are not performing their gender identity in an 'appropriate' heterosexually identified way. Similarly, many women who identify as heterosexual but who do not perform their gender in a way that can be read as differentiated from the opposite sex in a heterosexually desirable way also encounter harassment in the form of anti-lesbian abuse (Bunch 1991), whilst very 'feminine' lesbians can be taken for heterosexual. In these cases you don't have to *be* 'one' just to *look like* 'one' to be seen as a threat to the heterosexuality of the street. This just goes to show how the identity of those present in a space, and thus the identity of the space being produced, can sometimes be constructed by the gaze of others present rather than the performers.

Not all the processes at work maintaining the heterosexuality of the street directly involve violence and aggression. There are many other more subtle omni-present regulatory regimes constraining performative possibilities which pass unnoticed by those not subject to their pressures. Heterosexual looks of disapproval, whispers and stares are used to spread discomfort and make lesbians feel 'out of place' in everyday spaces. These in turn pressurize many women into policing their own desires and hence reinforce the appearance that 'normal' space is straight space (Valentine 1993). In this way, whilst sexual dissidents are constantly aware of the performative nature of identities and spaces, heterosexuals are often completely oblivious to this because they rarely have to be conscious of, or examine their own performativity. They can take the street for granted as a 'commonsense' heterosexual space precisely because they take for granted their freedom to perform their own identities. In contrast many dykes exercise constant self vigilance, policing their own dress, behaviour and desires to avoid confrontation. As Sally Munt describes:

There's nothing like being contained in its [Nottingham's] two large shopping malls on a Saturday morning to make one feel queer. Inside again, this pseudo-public space is sexualized as privately heterosexual. Displays of intimacy over the purchase of family-sized commodities are exchanges of gazes calculated to exclude. When the gaze turns, its intent is hostile: visual and verbal harassment make me avert my eyes. I don't loiter, ever, the surveillance is turned upon myself, as the panopticon imposes self vigilance.

(Munt 1995: 115)

As this quote neatly articulates, repetitive performances of hegemonic asymmetrical gender identities and heterosexual desires congeal over time to produce the appearance that the street is normally a heterosexual space. But, as the quote also demonstrates, this is an insecure appearance which has to be maintained through regulatory (including self-regulatory) regimes because the production of the heterosexual street is always under threat from sexual dissidents (re)negotiating the way everyday spaces are produced.

NOW YOU SEE US, NOW YOU DON'T: (RE)NEGOTIATIONS OF THE HETEROSEXUAL STREET

The identity of spaces and places, like the identities of individuals are 'frequently riven with internal tensions and conflicts' (Massey 1991: 276). Spaces are rarely being produced in a singular, uniform way as heterosexual (even though this is usually the hegemonic performance of space). Rather, as the quote above demonstrates, there are usually 'others' present who are producing their own relational spaces, or who are reading 'heterosexual' space against the grain – experiencing it differently.

'Lesbian desires and manners of being can restructure space' (Probyn 1995: 81) in many different ways. One of these is through dress. Subtle signifiers of lesbian identity, such as pinkie rings, labris earrings and rainbow ribbons; or lesbian dress codes such as butch-femme style, articulate subtly different spaces. Dress and body language (such as gestures, a glance, an independent or confident manner) also help dykes to 'spot' each other, to recognize a sense of sameness – she's like me – even though no words may be spoken. 'These features are not perceived or interpreted as indicators of lesbianism by straight women, because identifying lesbians is not relevant or necessary for them' (Painter 1981: 77). This is neatly captured by author Katherine Forrest in this excerpt from her lesbian crime novel *Amateur City*, when Ellen a gay woman caught up in a murder meets lesbian sleuth, Detective Kate Delafield for the first time:

> Ellen opened the door. 'Detective Delafield', she said.
> The woman sitting across from her at the conference table, her dark hair salted with gray, her corduroy jacket a light soft green, was examining a sketch, holding a leather-bound notebook sideways in strong square hands. She looked at Ellen with light blue eyes that were cool, level, and candid.
> Ellen stared at her. *Stephie can talk all she wants about not being able to tell for sure, but if this woman's not a lesbian then neither am I.*
> (Forrest 1984: 32, original emphasis)

Often it may be not so much what is there but what is missing, the wedding ring for example, that marks out (an)other identity. Thus through these fleeting glances or cruising, dykes can produce 'gay(ze) space' (Walker 1995: 75). Or as Probyn has argued 'space is sexed through the relational movements of one lesbian body to another' (Probyn 1995: 81). Sometimes the object of the gay(ze) doesn't reciprocate 'the look', rather a lesbian reading is imposed upon her, more in hope than anticipation. But the voyeur can still momentarily imagine the space as her own, producing a small fissure in hegemonic heterosexual space.

Lesbian spaces are also mobilized through linguistic structures of meaning. By 'dropping pins' for example, by referring to lesbian cultural icons or appropriated films, books or music, dykes can establish contact with other gay women, subverting heterosexual spaces with their own meaning.

> Lesbian social knowledge is used to interpret verbal and non-verbal behaviour [of others] such that the reality of one's lesbianism is not tied to external acts, but instead to the unquestioned and unquestionable propositions of the community itself. The woman, whether she perceives herself as a lesbian or not, is verbally constituted as a lesbian through the indexical use of members' talk.
>
> (Painter 1981: 72)

Hayes (1976) even goes so far as to suggest that there is a gay speech 'community' and gay lexicon – 'gayspeak'. Whilst language is undoubtedly important in constructing lesbian meanings, however, all sexual dissidents do not speak with one voice but are polyvocal.

Silence can also be a powerful way of articulating an identity: 'the lesbian, who does not engage in subtle verbal . . . patterns in straight settings to attract members of the opposite sex, notices or identifies other women who are also not engaging in these types of behaviour' (Painter 1981: 79) – and, again, can use these clues to establish a gay(ze) or relational space.

Like language, music also has the power to produce space. The music of artists such as kd lang and Melissa Etheridge, becomes infused with their lesbian sexuality, even though their lyrics and the sounds they make may have no explicit lesbian content and the artists themselves may resist this reading of their work (Valentine 1995). A lang track playing in street space, like a shop or a bar, can therefore facilitate the materialization of a lesbian space by causing two women to catch each other's eye and establish fleeting contact or even long-term friendships (Bradby 1993; Valentine 1995). 'Do you like kd lang?' is, after all, Cherry Smyth argues, *the* litmus test of a woman's sexuality.

Music also has the power to act as a vehicle that can transport the listener to another place and time. This use of music with fantasy allows women to use Walkmans to escape the street into imaginary lesbian spaces. Similarly, by playing music that has lesbian meaning publicly on tapedecks in shops or bars, or by using 'boom boxes' on the street, women can subvert straight space (Valentine 1995).

What these examples show is that:

> Lesbian identity is constructed in the temporal and linguistic mobilization of space, and as we move *through* space we imprint utopian and dystopian moments upon urban life. Our bodies are vital signs of this temporality and intersubjective location. In an instant, a freeze-frame, a lesbian is

occupying space as it occupies her. Space teems with . . . 'possibilities, positions, intersections, passages, detours, u-turns, dead-ends [and] one-way streets', it is never still.

(Munt 1995: 125)

However, these subtle possibilities and singular productions of relational sexed and gendered spaces often pass unnoticed by heterosexuals, either because overwhelming repetitive performances of heterosexuality swamp out these articulations of difference or because these subtle signifiers of 'otherness' are not read or understood by a heterosexual audience. Some lesbians therefore are actively using more 'in your face' tactics to challenge the stability of heterosexual productions of space.

'IN YOUR FACE': (RE)NEGOTIATIONS OF THE HETEROSEXUAL STREET

Don Mitchell has argued that public spaces are

very importantly, *spaces for representation*. That is, public space is a place within which a political movement can stake out the space that allows it to be *seen*. In public space, political organisations can represent themselves to a larger population. By claiming space in public, by creating public spaces, social groups themselves become public.

(Mitchell 1995: 115)

Lesbian and gay pride marches, held annually in Europe and the US, are one example of sexual dissidents being seen. By numerically appropriating the streets (and surrounding transport system, car parks, pubs, parks, shops, McDonalds and so on) and filling them with lesbian and gay meaning for one day, marchers pierce the complacency of heterosexual space. Sally Munt describes this spectacle as 'fifty thousand homosexuals parading through the city streets, of every type, presenting the Other of heterosexuality, from Gay Bankers to Gay Men's Chorus singing "It's Raining Men", a carnival image of space being permeated by its antithesis' (Munt 1995:123). But as this quote implies, Pride marches also achieve more than just visibility, they also challenge the production of everyday spaces as heterosexual. The importance of space is something that has particularly been seized on by queer activists.

In the early 1990s, impatient with lesbian and gay assimilationist tactics and inspired by aggressive unapologetic political tactics of AIDS activists such as ACT UP (AIDS Coalition To Unleash Power), a group met in New York to discuss how to resist the growing number of assaults against lesbians and gay men in East Village. Under the slogan of 'queers bash back', Queer Nation was born (Smyth 1992). This shift in politics (that quickly spread to Europe) away from integration and equality issues towards an 'in your face' confrontational tactics has also brought with it a recognition, to paraphrase Tim Davis, that heterosexism is a spatially constituted discourse that can be interrupted and undermined (Davis 1995: 287). Rather than merely trespassing in heterosexual public space with the political intention of staking out or gaining a share of it, queer is also about confronting and contesting the very production of public space.

Describing a hypothetical scenario of two lesbians in a pub kissing in front of a bar full of men, Elspeth Probyn explains how sexual dissidents can rearticulate space. She writes:

> while their kiss cannot undo the historicity of the ways in which men produce their space as the site of the production of gender (Woman) for another (men), the fact that a woman materialises another woman as the object of her desire does go some way in rearticulating that space. The enactment of desire here can begin to skewer the lines of force that seek to constitute women as Woman, as object of the masculine gaze . . . making out in straight space can be a turn-on, one articulation of desire that bends and queers a masculine place allowing for a momentarily sexed lesbian space.
>
> (Probyn 1995: 81)

One group that has set out not only to be seen, but also to bend and queer space, are the Lesbian Avengers. The notion of a lesbian direct action group was the brainchild of Sarah Schulman, an American writer and activist, who developed Lesbian Avengers with five friends, with the intention of prioritizing lesbians' needs and making lesbians visible (in both heterosexual and queer spaces). They began by recruiting activists at the New York Lesbian and Gay Pride where they circulated flyers stating: 'We want revenge and we want it now!' and with the invitation to 'Imagine what your lives could be. Aren't you ready to make it happen?' (Hopkins 1994: 18). This was rapidly followed by a series of actions to challenge heterosexism in everyday places, in which the Avengers sometimes joined forces with other groups, such as Las Buenas Amigas and African Ancestral Lesbians for Societal Change, to target specific organizations or places for protests (Meono-Picado 1995).

These tactics quickly spilled over into the UK. In July 1994 the birth of a UK version of the Lesbian Avengers was announced in the magazine *Rouge*: 'A new dyke direct action group has arrived, bringing with it style from AIDS activism, theatricality from queer and, above all, determination and bravado from suffragists, Greenham women and a long tradition of anger and action' (Hopkins 1994: 18).

The first action of the British group was to target the memorial to Queen Victoria (who famously denied that lesbians existed) near tourist focal point, Buckingham Palace. The statue was circled by over fifty women, chanting and banging drums and carrying banners such as 'Lie back and think of Lesbians' and 'The Lesbians are not amused' (*Diva* 1994). In September they followed up this action by taking over another space colonized with heterosexual meaning – the shop window (Reekie 1993). Invading window displays in one of London's major shopping streets, Oxford Street, lesbians posed next to the mannequins with labels such as 'Designer Dyke', 'Lesbian Boy' and 'Funky Femme'.

These 'in your face', aggressive tactics both raise the visibility of lesbians but also rupture the taken-for-granted heterosexuality of these spaces by disrupting the repetitive performances of the mall and the shopping street as heterosexual places and (re)imagining/(re)producing them as queer sites. Implicit in these 'other' performances of the street is a recognition that control over the way that space is produced is fundamental to heterosexuals' ability to reproduce their hegemony. The insecurity and anxiety of straights in the face of this challenge is evident in the heteropatriarchal jeers of passers-by and attempts to police such actions.

Disruptive performances of dissident sexualities on the street are therefore about empowerment and being 'in control'. These actions are also not only transgressive, in that they trespass on territory that is taken for granted as heterosexual, but also transformative, in that they publicly articulate sexualities (both lesbian and, by exposing its taken-for-granted presence in everyday spaces, heterosexuality) that are assumed to be 'private' (and in the case of lesbians also invisible) and thus change the way we understand space by exposing its performative nature and the artifice of the public/private dichotomy.

CONCLUSION: A KISS IS NOT JUST A KISS

The experience of the two Nottingham lesbians which opened this chapter demonstrates that a kiss is not just a kiss when it is performed by a same-sex couple in an everyday location. Rather, repetitive performances of hegemonic asymmetrical gender identities and heterosexual desires congeal over time to produce the appearance that street spaces (such as shops, parks and bars) are normally or naturally heterosexual spaces. The heterosexuality of the street is, however, as the response of the store manager to the lesbian embrace demonstrates, an insecure appearance that has to be maintained by regulatory regimes (from harassment and violence, to jeers and stares). It is insecure because space teems with many other possibilities. As Elspeth Probyn has argued 'we need to think about how space presses upon bodies differently; to realize the singularities of space that are produced as bodies press against space' (Probyn 1995: 83).

This chapter has tried to highlight two contrasting ways that lesbian bodies (re)produce space; first, by picking out some of the subtle ways women use lesbian manners and styles to fleetingly produce relational sexed and gendered spaces; second, by examining the more 'in your face' efforts of lesbian activists to visibly replace heterosexual space with other performances. These spaces are 'multiple and intersecting, provisional and shifting, and they require ever more intricate skills in cartography' (Rose 1993: 55) if we as geographers are to begin to try to map them.

ACKNOWLEDGEMENTS

I am grateful to Peter Jackson for his comments on an earlier draft of this chapter, David Bell for his legal advice and Nancy Duncan for her patience with my habit of missing deadlines.

NOTE

1 I'm using the term street in place of the word 'public space' to describe everyday publicly accessible places. This is because the term 'public' now seems inappropriate given, first, the way that many so-called 'public' spaces are now semi-privatized (i.e. are privately owned, controlled and managed); second, the fact these places are often not 'public' in that many people are excluded from them on the grounds of age, 'race', sexuality and so on; and third, because the term 'public' obscures the fact that many so-called 'private' relationships, such as sexualities, are actually part of 'public' space.

RENEGOTIATING THE SOCIAL/SEXUAL IDENTITIES OF PLACES

Gay communities as safe havens or sites of resistance?

Wayne D. Myslik

INTRODUCTION

It was late at night. I was walking to my car. At first they followed me in their car for a couple of blocks. I tried to lose them. Then they cut me off and got out of the car and chased me down. They kept screaming 'rich faggot'. I think there were four of them. They were wearing high school football jackets. They beat me up pretty bad.

(Don, 27)

I was walking down Connecticut Ave [near Dupont Circle] with another man . . . people driving down the street hanging out of the car screamed 'faggot' and threw a bottle . . . young high-school kids. A friend was physically beaten up outside the DC Eagle twice by groups of people who knocked him down. I know about a stabbing and drive-by shootings at Tracks.

(David, 37)

I was walking down the street at Dupont Circle, not doing anything overt. Two young white men in their twenties yelled out 'faggot'. A friend walked through P Street Beach [near Dupont Circle] and was assaulted. They broke his jaw.

(Glenn, 40)

For a two-year period you didn't go into [Dupont Circle] at night . . . There were serious bashings around the P Street area.

(Whit, 47)

A friend was assaulted and raped on the western edge of Dupont. You aren't as safe as you think you are.

(Peter, 25)

Most major American cities have neighbourhoods that are popular with gay men or that are predominantly gay. Exhibiting varying degrees of gay commercial and residential concentration, these areas offer opportunities for socialization with other gay men and provide access to specialized services ranging from bookstores to bars to clinics. In this chapter, I am particularly concerned with what I will call 'queer spaces'. These are areas that, being more than just concentrations of gay men, have come to be identified in and outside the gay community as gay spaces. By exhibiting a degree of social control by the gay community, queer spaces create the perception of being 'safe spaces'.

The concept of a 'safe space' is an important one for gay men, who are at risk of prejudice, discrimination and physical and verbal violence throughout their daily lives. Queer spaces are generally perceived as safe havens from this discrimination and violence, but they often serve as destinations of choice for 'gay bashers'. The experiences of Don, David, Glenn and Whit above refer to events in such queer spaces, near gay establishments, or near the homes of gay people. They portray the reality that incidents of heterosexist violence are common in these 'safe havens'. Why do gay men continue to identify queer spaces as safe spaces? To answer this question, we must consider the concepts of fear and safety and focus on the social contexts in which hetero-sexist violence, and reactions to such violence, take place. To do this, I will explore the following themes: (1) heterosexism as a cultural system and its enforcement, (2) violence against gay men, (3) issues of crime, safety, fear and vulnerability, (4) gay men's perceptions of straight places and queer places, and (5) power, politics, and territory as expressed in the landscape.

In approaching this issue from a geographical perspective, I intend to provide some insight into the role queer spaces play in helping gay men cope with the realities of heterosexism, and the violence that often accompanies it. This study discusses the experiences and perceptions of the white gay men who in many ways dominate the cultural landscape of the Dupont Circle neighbourhood of Washington, DC. Although many of these observations are not generalizable to other groups, they do raise important questions about how all of us perceive our places in the cities in which we live.

THE STUDY

Because gay men vary in the degree to which they are visible and there is, of course, no directory of gay men, it is impossible to know the size or location of the universe for this study. It was therefore impossible to use a true random sample. A quota sample was considered, but deemed unnecessary as the purpose of the study was to understand the individual experiences of each com-munity, not to find a representative sample of the entire city. More significantly, it would be necessary to designate quota sizes based on the entire population of the city (e.g. percentage of total city population that is working-class, etc.). It is highly questionable, however, whether the demographic characteristics of the gay population of the city mirror those of the greater city population, particularly by age.

The technique chosen for determining the characteristics of the sample, therefore, was judgement sampling, but individual sample units were identified through snowball sampling. Through my familiarity with the gay communi-ties of Washington, I had access to a significant number of individuals who were willing to assist in this study. These individuals, in turn, were asked to identify others like themselves who would be interested in participating in the survey. A particular concern in this process was to avoid the dangers of choosing a sample made exclusively of activists or otherwise extroverted individuals who would be unusually inclined to volunteer for such a survey. In order to obtain as broad a sample as possible, I placed advertisements in the city's gay news-paper (*The Washington Blade*), the city's free newspaper (*The City Paper*) and the neighbourhood paper (*The InTowner*). I also posted flyers at local supermarkets, bookstores, bars, discos, restaurants, coffee houses, video stores and sex shops frequented by gay clientele.

The final sample included fifty gay white men between the ages of 23 and 48. Approximately 75 per cent of them identified themselves as middle class, the remaining 25 per cent as working class or 'blue collar'. Approximately 70 per cent were living in Dupont Circle or in the process of moving into the neighbourhood from elsewhere in the city. The remaining 30 per cent worked in the neighbourhood or spent all of their free time there. The interviews, although based on a survey guide, took the form of open conversations lasting upwards of one and a half hours each. Most interviews were conducted at a restaurant or coffee house chosen by the subject. A few were done at the sub-ject's home or over the phone, as they felt most comfortable. Approximately 30 per cent of the subjects contacted me on the recommendation of a friend who had already been interviewed; 30 per cent responded to advertisements and flyers; 30 per cent were contacts I had made socially or professionally while living previously in the Dupont Circle neighbourhood.

HETEROSEXISM

It's so oppressively straight there [Georgetown].[1] If feel out of my element. They don't even conceive of it . . . that gay people might exist. I'm scared, I guess.

(Geoffrey, 27)

I cease anything that may be construed as gay behaviour. You have to act asexual.

(Ron, 23)

I'm supposed to act a certain way. I'm only tolerated within certain parameters.

(Frank, 37)

I always feel like when I walk through a het crowd, maybe someone can tell something is different. I can't relax for fear that I will get a verbal attack, a strange or awkward look.

(David, 37)

The system

Feminist theorists have demonstrated the degree to which gender relations are reflected in and constitutive of the patriarchal organization of space in

Western culture (see, for example, England 1991; McDowell 1983). Intersecting this gender-ordered construction of society is the sexual ordering of space and place under what Valentine has called 'heteropatriarchy, that is, a process of sociosexual power relations which reflects and reproduces male dominance' (1993: 396).

Western society is based on the notion that the natural purpose of sexuality is for reproduction and that sexual identity is linked inextricably to the individual's role in the reproductive family. Organized around the construction of heterosexuality as the dominant and 'normal' form of sexual identity, this view of sexuality is directly dependent upon a binary system of masculine and feminine gender identities that are believed to coincide directly with male- and female-sexed bodies. Gay men thus become outlaws, alien to this heteropatriarchal system.

The current use of the term 'family values' by the religious right in attacks against gay civil rights is a cogent example of the extent to which a heteropatriarchal society sees gay men as alien to the familial system. Sexual relations between gay lovers are illegal in many areas, gay marriages are not recognized and the courts frequently deny the rights of gay men to be parents by taking their children from them. With vicious irony, the religious right then labels gays as 'anti-family'.

Providing abundant evidence of the gendering of urban public space, feminist geographers have been very successful in breaking down the traditional distinction between public and private space that sees sexuality as limited to the private domain (see Duncan, this volume; Peake 1993; Valentine 1993). They have shown that all space has a gender identity and that most spaces, public and private, are masculine dominated. Just as spaces may be identified as masculine or male-dominated, though, urban spaces also have a sexual identity. Virtually all such space is heterosexually dominated. As Davis has pointed out, however, 'heterosexism and homophobia are social constructions with spatial impacts that arc not always clearly visible in the physical landscape' (1994: 2). Indeed, as Geoffrey notes when he says 'they don't even conceive of it . . . that gay people might exist', most people are blissfully ignorant of the degree to which sexuality, and in particularly, heterosexuality, permeates space. Illustrative of this ignorance is the often heard statement that gays would be tolerated if they didn't 'flaunt' their homosexuality. Inherent in this statement is the assumption that heterosexuality is itself not flaunted or expressed outside the home. However, engagement announcements, bridal showers, wedding ceremonies and rings, joint tax returns, booking a double bed at a hotel, shopping together for a new mattress, casual references in conversation to a husband or wife, a brief peck on the cheek when greeting or leaving a spouse, photos of spouses on desks at work, holding hands at the beach, and even divorces are all public announcements and affirmations of one's heterosexuality. The 'normality' of heterosexuality is so deeply ingrained in Western culture that it is not even seen. Gay men, though, are keenly aware of this 'heterosexual assumption', its visibility, and its impact on their lives. As Ron and Frank have noticed, gay men are tolerated only so long as their gay identity, their homosexuality, remains hidden. When it becomes visible, one is at risk of attack. A t-shirt popular among many gay men states, 'I don't mind straight people, as long as they act gay in public.'

In nearly all public spaces, then, there is no tolerance for departure from a heterosexual gender-identity and its attendant patterns of behaviour. Gay men learn that in the workplace, in bars, in shopping malls, on the street, in virtually every physical or social space in which they travel, sexual orientation must never be visible. For most gay men, adapting behaviour between gay and straight spaces to hide their sexual identity becomes natural and nearly unconscious. They do it as automatically as other men change their behaviour when they take off a sweatshirt and put on a business suit.

The enforcement

According to Gary Comstock's extensive survey of male victims of anti-gay violence, perpetrators are typically white (67 per cent) males (99 per cent) under 21 years old (50 per cent) and outnumber their victims. This observation is supported by the survey I conducted of gay men in Washington. Of those who were victimized and knew the race and age of the perpetrator, nearly all identified their attackers as white, male, teenaged or early twenties. All were outnumbered by their attackers. It is perhaps most significant that perpetrators of anti-gay violence do not typically exhibit expected criminal attitudes and behaviours. They rarely have the histories of criminal activity or psychological disorders typical of other violent criminals. It is commonly observed, by victims as well as defence attorneys, that perpetrators of anti-gay violence are 'average boys exhibiting typical behavior' (Comstock 1991: 93).

The explanation of how such violent behaviour can be deemed 'typical' must be found in the socialization of men in American society. Men are socialized to be dominant and aggressive, to conform strongly to established sex roles and to ridicule or punish those who deviate from those roles. Most American men continue to identify gay men and women according to stereotypes of the effeminate 'sissy' or the masculine 'bull dyke'. This association of homosexuality with deviation from accepted gender roles and violation of mandatory heterosexuality interacts with the gender-role socialization of men. Gay men thus become identified as a group requiring ridicule, policing and/or punishment.

Furthermore, the role of adolescence in modern Western society interacts with male socialization to produce a group of people with a greater social incentive to target gay men. Since the nineteenth century, adolescence has become a 'kind of temporal warehouse or greenhouse in which young people are parked until needed' (Comstock 1991: 103). Here they are forced to wait for social maturity, before they are told that they may fully participate in society in the roles of wife, husband, parent or worker. Although at their sexual prime, adolescents are told that they are not ready to take on the responsibilities of family. Furthermore, they are forced to compete for boring, part-time and temporary work in the service sector. 'Ready to develop primary and independent affective relationships and to take on occupational challenges, they remain their parents' children, dependents with minimal earning power' (Comstock 1991: 103).

Placed in this frustrating position, denied any sense of power or control over their lives, but encouraged to anticipate such control and responsibility in the future, it is hardly surprising that adolescents should resort to activities and behaviours that strengthen and affirm their social status. The socially constructed powerlessness of adolescents, then, is a partial cause of problem behaviour, which can be understood as identity-building and power-seeking at the expense of others who also lack power in the social order. Such adolescents can affirm their power only over others with similar or lower status in society. Their targets, therefore, are members of groups shunned and denigrated by adults. Their violent actions are perversions and exaggerations of that adult behaviour.

Gay men thus serve as a fixed low-status standard away from which adolescents move up in social standing. Those with the highest social expectations, white males who are held back from high-status positions they are told they deserve by virtue of gender and race, are most likely to express frustration and demonstrate power over lower-status groups. It is not surprising, then, that middle-class adolescents, with the highest expectations for the future, are disproportionately represented among the perpetrators of heterosexist violence. It would be particularly interesting to see what percentage of these adolescent males also commit hate-based crimes against groups such as blacks and Jews, or develop histories of violence against women.

These behaviours are often ignored or even condoned by parents and reinforced by the institutions of society. Churches preach that homosexuality is a sin, schools fail to protect gay youths from harassment, police inadequately respond to assaults, and the courts give perpetrators light sentences. Victims are often blamed for their attacks, and often suffer secondary victimization at the hands of police, medical and judicial officials (Anderson 1982: 146). It is well known that the majority of incidents of heterosexual violence go unreported, the victims often fearing further abuse from the justice system (Herek and Berrill 1992: 289). Most of the men who spoke to me in Washington expressed a cautious optimism about relations between the police and the gay community. Several believed the situation has improved in the past few years. However, even the most optimistic men who spoke to me are reluctant to give the police the benefit of the doubt. Incidents of discrimination or ill treatment by the police are quickly reported throughout the community and are not forgotten for years. Although the police are seen as potentially helpful, it would be unsafe to trust or rely upon them.

We can thus see most incidents of anti-gay violence as the acts of young men attempting to affirm their individual status within their peer group and their group status in society. They accomplish this by demonstrating their adherence to gender roles and asserting social and sexual dominance over a lower-status group whose marginalization by society identifies them as acceptable targets. Attacks on such groups are inadequately proscribed by society and may even be rewarded. These acts of violence are not personal expressions of intolerance of homosexuality, but of societal intolerance, or cultural heterosexism, which grants permission for their actions and mitigates their responsibility for the consequences. Heterosexist violence, therefore, like violence against women, must be understood in the broader context of social dominance based within the structures of privilege derived from race, gender and sexuality.

The violence

Since the Second World War, gay men have made significant progress in increasing their visibility in American society and in creating organizations that are fighting for their civil rights. Since the Stonewall Riots of 1969, gay social, cultural and political organizations have appeared in nearly every major city, moving out of rented halls and private homes into permanent office spaces and purchased buildings.[2] This creation of visible queer spaces has been met with a drastic increase in violence against gay men and their property. In their comparison of thirteen major surveys of anti-gay violence, Herek and Berrill found that 80 per cent of gay men and women surveyed had been verbally harassed, 44 per cent were threatened with violence, 33 per cent had been chased or followed, 25 per cent were pelted with objects and 19 per cent were physically assaulted (Berrill 1992: 26). Another study found that gay men are at least four times as likely as the general population to be violently attacked (Comstock 1991: 55).

A report put out by the National Gay and Lesbian Task Force in early 1994 reports that incidents of violence against gay men increased by 127 per cent between 1988 and 1993. This study has suggested that, for the first time since such surveys have been conducted, reports of certain types of anti-gay incidents have decreased (NGLTF 1994). Although the NGLTF has been wary of drawing conclusions from this study, it would appear that a pattern of hot-spots is developing. Reports of minor harassment have declined, while reports of physical violence are increasing in certain local communities, particularly in those where gay rights issues are being confronted. In Washington, DC, the organization Gays and Lesbians Opposing Violence has issued a report in response to the NGLTF survey which indicates that violence in the District of Columbia continues to increase.

Queer spaces, those areas in which gay men are known to congregate and have been designated as safe spaces, are, ironically, the most frequent settings for this violence, exceeding by 28 per cent straight public areas. As an area becomes more generally recognized as a gay neighbourhood or queer space, the violence increases and perpetrators travel in search of their targets. The Castro District in San Francisco illustrates this point well. Statistics gathered by the grassroots organization Communities United Against Violence showed that in the 1970s 93 per cent of reported assailants were youths from the neighbouring Mission and Filmore districts. By 1981, when the Castro had a well-established reputation as a gay neighbourhood, the percentage of perpetrators who were travelling from more distant areas had more than doubled (Comstock 1991: 61). It is clear that as the reputation spreads in the media that an area is a gay neighbourhood or queer space, the violence increases. Queer spaces become hunting grounds.

CRIME AND SAFETY

I fully expect to be harassed sometime.

(Robert, 32)

I expect that as long as I live there will be homophobes trying to outmanoeuvre me.

(Scott, 42)

I don't think this problem is going to get any better . . . There are going to be more and more gangs and gay bashing.

(Rod, 28)

I always have the feeling that if I were recognized as gay, chances are it would happen.

(Bob, 44)

I'm afraid whenever I walk down an unfamiliar street and I don't see any other gay people.

(Frank, 37)

Gay men are not unaware of their risk of victimization. In a national survey conducted in 1984, 83 per cent of gay men said they believe they might become the victim of heterosexist violence in the future (Comstock 1991). In my survey of gay men in the Dupont Circle neighbourhood of Washington, DC, nearly 85 per cent said they expect to be victimized in the future. The perception of vulnerability expressed in the quotations above can affect an individual in many different ways. It can be related to fear, anxiety and stress which can affect the personal and professional life. It can lead to editing behaviour, forgoing activities and severely limiting the quality of life. Along with the effect on an individual's quality of life, limiting behaviour in response to such crime serves to perpetuate the invisibility of gay men, strengthening the heterosexist and homophobic system that gives rise to the violence in the first place.

Fear, vulnerability and behaviour

Since the early 1970s, several disciplines including criminology, sociology, psychology and geography have recognized that fear of crime is an important social problem and that numerous social, spatial and psychological factors contribute to the phenomenon. Understanding in particular that fear of crime is only loosely linked to the real presence of crime in an area, most studies since the late 1970s have focused on the independence of fear from victimization. The empirical research has typically examined communities experiencing socioeconomic decline, characterized by a low rate of crime but a very high fear of crime.

Few studies, however, have examined the inverse of this relationship, in communities where there is a high crime rate of high expectation of crime but low level of fear. It is not difficult to imagine how a community with a low crime rate can nonetheless come to experience fear. However, it seems counter-intuitive that a community with a high crime rate, in which individuals expect to be victimized, could be characterized by the absence of any fear or anxiety and the presence of a strong sense of safety. I found just such a counter-intuitive relationship in the gay community of Dupont Circle.

Most early studies of fear of crime relied on national or city-wide crime surveys, did little to distinguish between fear of different types of crime and were generally concerned with determining the predictive value of certain demographic characteristics, such as age, sex or income. These studies provide little explanatory insight. Studies by social psychologists are more useful in explaining the relationship between perceptions of crime, fear and precautionary behaviour (see, for example, van der Wurff 1989). However, such studies rarely have looked beyond individual cognitive processes to place fear of crime in a broader societal context.

Geographers and their spatially minded colleagues have understood for some time that the physical and social environments interact to produce certain perceptions of crime or safety. This view, referred to as the disorder model, is described succinctly by Bursik and Grasmick, 'Fear is a response to the perception that the area is becoming characterized by a growing number of signs of disorder and incivility . . . that indicate that the social order of the neighborhood is eroding' (1991: 101). Residents cannot be assured that others will adhere to a shared set of expectations about behaviour. At least three broad aspects of the social environment can be identified as contributing to fear of crime relatively independently of the experience of victimization. The first is environmental incivility, in the form of abandoned buildings, vandalism and graffiti; the second is a lack of community spirit or community satisfaction; and the third is racial tension, most often the result of changes in the racial composition of the area (Smith 1986).

More recently, feminist researchers have created a significant literature on women's fear of crime which analyses the position of women in society and its significance for fear of crime (see, for example, Pain 1991; Riger 1991; Valentine 1989). These studies recognize that women are subject to a form of violent crime, rape, which is rarely a concern for men. Women express strong feelings of vulnerability to rape because of the likelihood of serious injury in addition to the rape, the perception that they could not physically defend themselves and the lack of protection they receive from society.

Few of these studies, however, have been able to separate perception of crime, feelings of vulnerability, fear and behaviour changes from each other. Although difficult to discern from a study of a group that experiences high levels of vulnerability, fear and behaviour modification, this distinction is significant. Not all groups are socialized to experience or express emotions such as fear in the same way. For example, many men will recognize that crime is a hazard in their neighbourhood and take steps to prevent their victimization, but they will be loath to admit that they experience fear, much less that it is a significant motivation for their behaviour. Women, however, are more likely to identify fear as a motivating factor. Furthermore, many men are not even consciously aware of the behaviour changes they do make. When these changes are observed, many men will attribute them to 'common sense' rather than a reaction to fear or concern for crime. It is misleading, therefore, to rely on the term 'fear' in discussion of crime perception. It is more accurate to discuss the breakdown of feelings of safety or security. Their absence from a neighbourhood can be expected to reflect the presence of a general sense of safety in the community.

Straight places

Like women, gay men are confronted with a particularly dangerous crime, bashing, which is associated with high levels of injury but little societal protection. Gay men's perception of their risk is particularly high and, like women, they report feelings of vulnerability in most parts of the city. Gay men, however, are men. As Robert, Scott, Rod, Bob and Frank illustrate, they express their concern in many ways, but few will admit feeling fear. Had I limited my question to 'Are you ever afraid . . . ' as many surveys do, it is unlikely I would have discovered the degree of concern shown by these men.

> I don't act any differently . . . No, I wouldn't show affection to a man. Yes, I would in a gay place. I guess I do act differently!
>
> (Bob, 44)

> I edit a small amount of my behaviour.
>
> (Glenn, 40)

> I am less leisurely, more businesslike. I avoid touching other men or 'showing queer'.
>
> (Basil, 44)

> I don't hold hands. I just don't even think about it.
>
> (Ron, 23)

> I never show affection.
>
> (John, 48)

Like straight men, gay men frequently are not conscious of the limits they place on their behaviour, and are likely to claim such behaviour is 'common sense', or 'just the way things are', rather than attributing it to concern over crime. When asked in general terms if they change their behaviour at all, nearly 60 per cent told me they do not. When asked about specific behaviours, such as 'Do you hold your lover's hand when you are out at a gay bar, or on the street in Dupont Circle?' and 'Do you ever hold your lover's hand when you are in a restaurant or on the street in Georgetown?' they became aware of their behaviour. Outside the gay neighbourhood or gay establishments, more than 80 per cent avoid shows of affection, physical contact with other men, speech patterns or vocabulary considered stereotypical, and clothes or other signs that they are gay. Because this editing of behaviour has been the norm for most of these men's lives (their openness around other gays is the exception), it is usually unconscious, or associated with 'normal' behaviour in certain circumstances. Often they commented that Georgetown, for instance, just is not a place men show affection. They seemed unaware they were changing their behaviour or why. As Bob's statement above illustrates, during the interview, many of men realized for the first time what they were doing.

Queer places

I feel safer . . . around people tolerant of my identity.

(Frank, 37)

I'm not preoccupied with the place I'm in or how I'm being perceived.

(Charles, 24)

Most gay men who spoke to me realize that the danger of heterosexist violence is greatest in queer spaces. However, all the men who spoke to me said they feel safer in Dupont Circle than in any other part of the city. What is so special about queer spaces? Why do gay men identify the most dangerous neighbourhood as their 'safe space'? To answer this question, we must understand how feelings of vulnerability, fear and safety are related to feelings of social control and to the construction of power and privilege in society.

Looking for signs of disorder which might be related to feelings of safety or security in Dupont Circle, one finds that homes and businesses are well maintained and there are no abandoned buildings or vandalism. Furthermore, the area is, on the whole, racially homogeneous. Not surprisingly, then, one does not find a high degree of fear over general crime.

Of particular importance to the gay community and its attitudes toward heterosexist violence, one finds a very strong sense of community spirit in Dupont Circle. Of the men I interviewed who live in the area, over 80 per cent identified a desire to be among other gay men as a major factor in their choice of neighbourhood. Of those who do not live in the area, more than half said their next move would be to Dupont Circle for these reasons. It is to visit gay friends and partake in this community spirit that most non-resident gays come to the neighbourhood. Moreover, although one often finds graffiti on the sidewalk or lamp-posts, they are usually pro-gay statements by an activist group such as Queer Nation or Act Up. These graffiti reinforce a sense of community territory among gay residents.

Despite the problem of anti-gay violence and harassment, Dupont Circle maintains a sense of social order and control which is partly responsible for feelings of general safety. However, one must remember that nearly all the men surveyed expect to be victimized by heterosexist violence, report general feelings of vulnerability to such crimes and are aware that these crimes are most frequent in this neighbourhood. But all feel safest in this neighbourhood. No matter how clean the streets or how well maintained the homes, the simple lack of signs of disorder is not enough to account for such a contradiction in attitudes. To understand the magnitude of gay men's sense of safety and control, one must understand the broader context of social power, what it means to be gay in the city, and what it means to be a gay man in a queer place.

POWER

You come up the escalator out of that dark [metro] tunnel at Q Street, and you see where you are, and there's such a sense of . . . empowerment.

(George, 25)

It was like stepping out of the closet!

(Don, 27)

As George described how he feels when he visits Dupont Circle his face lit up and his whole body became animated. His trip out of the dark metro tunnel had meaning for many of the gay men with whom I spoke, such as Don, who also was describing the first time he came out of that dark metro tunnel into the sunlight. For these men, Dupont Circle is more than just a residential or commercial concentration of gay men. It is a political and cultural centre with great emotional significance.

Politics

For over twenty-five years, the gay neighbourhood has served as the gay community's centre for political activity and electoral power. Davis, however, provides evidence that the day of queer space as the centre of political influence for gay men has passed, for

> the power to create real social and political change . . . has been dispersed in such a manner as to make it impossible to create wholesale change through progressive legislation . . . Legislative victories are increasingly symbolic, when real acceptance can only be created in the cultural sphere.

(Davis 1994: 1)

In Davis's estimation, the gay territory has failed in its mission of creating a 'liberated zone' and has become the 'gay ghetto', a symbol of 'isolation and continued oppression' (Davis 1994: 1). While his assessment of the effect on gay politics of the changing locus of political power in postmodern society is an astute one, I do not agree with his pessimistic view of the new role of the 'gay ghetto'. Although the development of a local base for electoral power was a significant effect of the claiming of urban territory by gay men, the earliest, most important, and continuing roles of these queer spaces have been social, emotional, cultural and symbolic, not overtly political.

Moreover, even as the locus of political power moves from the legislative to the cultural sphere and the significance of gay neighbourhoods to local political struggles decreases, the influence of queer spaces continues to rise. As sites of cultural resistance with enormous symbolic meaning for gay men, such spaces provide cultural and emotional support for a political movement comprised of an increasingly diverse and geographically scattered community.

Nearly fifteen years ago, John D'Emilio observed that, 'for gay men and for lesbians, San Francisco has become akin to what Rome is for Catholics: a lot of us live there and many more make the pilgrimage' (D'Emilio 1981: 77). I suggest that today, queer spaces such as San Francisco are for gay men more akin to what Jerusalem is for Jews: most of us live somewhere else, fewer of us make the pilgrimage than in the past, our political power has moved elsewhere, but the cultural and emotional significance of the place cannot be overestimated. For those who do choose to live in or use them, queer spaces represent, if not a physically safe 'liberated zone', a site of cultural resistance where one can overcome, though never ignore, the fear of heterosexism and homophobia.

Territory

This is our territory, an area we've claimed.

(Stuart, 26)

When asked what it means to be 'safe' as a gay man, nearly half of the men who spoke to me defined safety not in physical terms or in terms of violence and crime, but as 'living openly as a gay person', 'being comfortable in my sexuality'. Many gay men explain that they feel safe in queer spaces because of a sense of safety in numbers. Interestingly, though, these men do not believe that other gay men are more likely than straights to come to the assistance of a bashing victim. In fact, several suggested that gay men, out of fear for their own safety, might be less likely to intervene in an attack or bashing. The safety they feel, therefore, is clearly an emotional and psychological safety that comes from being in an area in which one has some sense of belonging or social control, even in the absence of physical control.

Dupont Circle is territory that has been claimed by the gay men of Washington. Although there are violent exceptions, the residents of Dupont Circle have come to expect others to adhere to certain behavioural patterns, the most significant of which is a tolerance of open expressions of homosexuality and the open association of gay men. When Stuart states, 'This is our territory', he is not only attesting to his claim on the Dupont Circle neighbourhood, but reaffirming that claim. For gay individuals, alienated and with no sense of power, control or order in any other part of the city, this sense of claimed territory takes on an enormous emotional significance.

CONCLUSION

Queer spaces are not, in fact, safe havens from the threat of violence that follows gay men throughout their lives. Ironically, the congregation of people which provides an emotional and psychological safety itself undermines physical safety by advertising the existence and location of a target group. 'Safe' spaces in turn become hunting grounds.

However, despite the evidence that gay men are especially targeted in these areas, queer spaces in many respects alter the traditional power relationship between heterosexuals and homosexuals. In most areas of the city, gay men feel uncomfortable or vulnerable and, consciously or unconsciously, alter their behaviour to hide their sexual identity. As sites of resistance to the oppressions of a heterosexist and homophobic society, however, queer spaces create the strong sense of empowerment that allows men to look past the dangers of being gay in the city and to feel safe and at home. Overwhelmingly, they consider the psychological and social benefits of open association worth the physical risk taken in queer spaces. For gay men, coping with the presence of violence is an act of negotiating power in society.

NOTES

1 Georgetown is a wealthy neighbourhood near Dupont Circle which is home to Georgetown University and a strip of shops and bars crowded with presumably straight college students. Owing to its palpably heterosexual atmosphere and the numbers of drunken young men, many gay men list Georgetown as the neighbourhood they feel least safe.
2 On 26 June 1969 a police raid on the Stonewall Inn, a gay bar in New York City, set off a riot that is considered by many as the symbolic beginning of the modern gay rights movement.

ON BEING NOT EVEN ANYWHERE NEAR 'THE PROJECT'

Ways of putting ourselves in the picture

Vera Chouinard and Ali Grant

INTRODUCTION:
SO WHAT IS 'THE PROJECT' ANYWAY?

The last few years have seen massive upheaval in progressive theory, research and politics (Faludi 1992; Fraser 1989; Harvey 1989; Palmer 1990; Ross 1988). In geography, as in the social sciences more generally, there have been subtle and complex redrawings of 'The Project' (Barrett and Phillips 1992; Christopherson 1989; Fraser 1989; Walker 1990). For radical geographers (in which we include feminist geographers), these changes have been signalled by the appearance of collections showcasing new critical geographies (e.g. Kobayashi and MacKenzie 1989; Peet and Thrift 1989; Wolch and Dear 1989) and gradual shifts in concepts, research foci and methods. Put simply, critical geographers have engaged in a partial retreat from class analysis and issues and increasingly recognized the value of feminist geography (although the terms of 'engagement' or negotiation very much remain to be worked out). Marxist and feminist geographers have also been searching for theories and methods which recognize diversity in human lives and in the meanings assigned to those lives. This has led to some serious and not-so-serious flirtations with postmodern philosophy and theory (Dear 1988; Deutsche 1991; Harvey 1989; Soja 1989).

What, then, is the new 'Project' in radical geography? It would be misleading to suggest that we can outline all its contours, or provide a single definition that would satisfy all radical geographers. We can, however, zero in on some of its most central features – and it is important to do so, because in some ways 'The Project' is as partial and exclusionary as those that have gone before.

One key feature is an emphasis on processes of oppression rooted in three sets of relations: class, gender and race. In conference sessions we hear this 'trinity' so often it has almost become a group chant, and the radical geographic literature increasingly flags these relations as sources of oppression. At one level, of course, these developments are wonderful and exciting: radical geographers have fought long and hard to have the significance of gender and race, as well as class, recognized. At another level, however, through their silences, they signal important exclusions in 'The Project'. The disabled, lesbians, gays, the elderly and children are frequently invisible in definitions of 'The Project' – definitions which fail to acknowledge ableism, heterosexism and ageism as significant sources and structures of oppression. Notice as well that we radical geographers seldom discuss the politics of research practices as part and parcel of our efforts to reconstruct 'The Project'. If we did so more often, perhaps these silences could be effectively challenged. Perhaps this is exactly why we don't.

In what follows, we take a critical look at two of these silences and discuss what geographers might do about them. We look at 'The Project' from two perspectives: those of lesbian and disabled women. From these vantage points, it becomes clear that it is high time for all geographers to do what they can to ensure that 'other' voices and practices are taken seriously in struggles to reconstruct 'The Project'. This does not mean simply tinkering with theories and methods so that lesbians and disabled women become another 'topic' or 'viewpoint'; it means drawing on these vantage points in revolutionary ways which challenge and disrupt our understanding of processes of exploitation and oppression. It means seriously confronting the possibility, indeed the likelihood, of our own complicity in the construction and oppression of 'others', and struggling to find ways to understand these processes. It means not being content with representations of inclusion, but insisting that the voice and presence of 'others' in the research process is an essential part of the struggle for social change.

Our own politics and experiences presented challenges in co-authoring this paper (see also Bell *et al.* 1994). Although we have common experiences, and are therefore able to talk about each other's oppression *in general*, neither of us felt it was possible or appropriate to represent the other's *particular* oppression. Thus one of us wrote the sections on disability and ableism, the other the sections on lesbian experience and heterosexism. We argued over each other's sections and did not always agree; this was not a problem. Language did present a problem: using 'we' throughout the paper would have implied that we both felt equally able to discuss each other's oppressions, but how could we switch from 'we' to 'I' without thoroughly confusing the reader? There is no easy answer, so 'I' in the discussions of disability belongs to the first author, and 'I' in the discussions of lesbian experience belongs to the second.

LIVING EXCLUSION:
TWO GEOGRAPHERS' TALES

This section discusses what it is like to live in an ableist and heterosexist society. Ableism is defined here as any social relations, practices and ideas which presume that all people are able-bodied. Examples include: evaluating disabled workers by the same criteria used to evaluate able-bodied employees, holding events in physically inaccessible locations and treating not being able-bodied as defining a disabled person. Heterosexism refers to social relations, practices and ideas which work to construct heterosexuality as the only true, 'natural' sexuality whilst negating all other sexualities as deviant and 'unnatural'. Examples include: legal definitions of 'family' which do not include same-sex couples, assumptions that peoples' partners must be of the opposite sex, hostility toward lesbians and gays who make themselves visible in territory dominated by heterosexual relations and norms (e.g. public places and workplaces), and failing to recognize and appreciate lesbian and gay cultures.

We begin with personal experiences to help the reader 'see' society and our discipline through others' eyes. The intention is not to evoke sympathy or pity but to encourage the reader to understand more fully the environments we both negotiate daily. While we present experiences of ableism and heterosexism separately, this should by no means suggest that these are separate from oppressions based on class, gender, race, ethnicity and age. Rather, there are multiple locations of oppression in patriarchal, capitalist societies including those based on both ableism and heterosexism.

The disabled woman

It is hard to think of any facet of my life which has been untouched by ableism and by struggles to occupy able-bodied spaces on my own terms.

For example, as a professor, my workplace, the university, has been a very significant site of my oppression. This has taken multiple forms, from physical barriers to access to the use of ableist standards to evaluate my contributions. One form of exclusion is very visible: after four years, I still lack physical access to my office. Two entrances which appeared as scooter and wheelchair accessible on our official map for disabled staff and students turned out to be nothing of the sort: there are no automatic doors, no working lift and heavy internal fire doors block access to corridors. So, although I have acquired a scooter, I still cannot get into my official workplace independently. This situation sends out strong signals that my presence, and the presence of disabled colleagues and students, is not important; that we are not valued in the academic setting.

Ableist relations and practices are manifested in a number of other ways as well. For instance, there is no procedure to adjust the workload required of disabled professors, although a full workload for many disabled people should be defined not in terms of work expected from able-bodied professors, but in terms of the capacity of the disabled academic. Other social barriers include a reluctance to be flexible in terms of how classes are taught (e.g. in the disabled professor's home or through interactive computer technology).

To these social manifestations of 'ableism' one must add the little everyday practices of academic life which exacerbate the challenges of being disabled. Recently, in a feminist geography conference session, I was forced to stand because the room was filled. This was arguably my own 'fault' as I arrived late (having had to walk a long distance from another session), but after about half an hour the pain in my feet, legs and hips was so intense that I was forced to ask a young woman if she and her companions could shift one chair over so that I could sit down (someone had left their seat, so there was an empty chair at the far end of the row). I apologized for asking, but explained that I was ill, very tired and in a lot of pain. She turned, looked very coldly at me and simply said 'No, the seats are being used.' She may well have been right, but I suddenly no longer felt part of a feminist geography session: I was invisible . . . and I was angry. Fighting a juvenile urge to bop her on the head with my cane, I began to see feminist geography through new eyes; eyes which recognized that the pain of being 'the other' was far deeper and more complete than I ever imagined, and that words of inclusion were simply not enough.

Other negative reactions to my being disabled include direct challenges to my right to occupy able-bodied territory. Recently an older woman burst into a shop where I was sitting chatting with the owner (a friend of mine). Sticking her face uncomfortably close to mine (invading my territory!) she looked at my arm braces and walker and blurted out 'My god, what the hell happened to you?' I explained that I was disabled by rheumatoid arthritis. 'Oh', she replied, and walked away. She had no interest in my well-being; she was simply asserting her 'right' as an able-bodied woman to demand explanations for the presence of the disabled in 'her' space.

My sense of myself, as a disabled, academic woman has also been shaped by more subtle aspects of daily life. Walking on the university campus and in other public places, I am constantly conscious of frequent looks (often double and even triple-takes). I realize it is unusual to see a relatively young woman walking slowly with a cane or using a scooter and the looks reflect curiosity, but they are a constant reminder that I am different, that I don't 'belong'. It is painful for me to acknowledge this. I guess that is why I have learned to look away: to the ground, to the side . . . anywhere that lets me avoid facing up to being the 'other'.

It is remarkable how thoroughly ableist assumptions and practices permeate every facet of our lives, even though we often remain relatively sheltered from and insensitive to these forms of oppression. Yet disability in some form will come to each and every one of us someday, and when it does, and ableism rears it ugly head, one finds a topsy-turvy world in which none of the old rules apply and many 'new' ones don't make sense. People develop new ways of relating to you often without recognizing it. For instance some of my students will not call me at home, despite instructions to do so, because I am 'sick'. Other students shy away from working with a disabled professor: some assume that the best, most 'successful' supervisors must be able-bodied; others are unwilling to accommodate illness by, for example, occasionally substituting phone calls for face-to-face meetings or meeting at my home rather than the office. Of course this is not true of all students, but these practices are pervasive enough to hurt every day to make it just a little harder to struggle to change relations, policies, practices and attitudes.

Some manifestations of ableism would be hilarious if they weren't so hurtful and damaging. An administrator heading a university disability programme invited me to sit on a committee planning events for our annual disability awareness week. I explained that I would be happy to contribute but, as I was quite ill and immobile at the moment, we might at times have to settle for a phone conversation rather than my actually attending a meeting. The response was absolute silence, even though I was told those responsible would get in touch. It was a shock to realize that even those in charge of disability issues could act in such ways – a bit like being dropped in the middle of the Mad Hatter's tea party without being told what story you were in.

The dyke

Like disabled women, lesbians must struggle over the right to space, in a culture that constructs 'woman' as both able-bodied and heterosexual. And as a lesbian who tries to be always 'out' in everyday spaces, I spend most of my time fighting for some space, whether in a geography department, in a restaurant or on the street. Heterosexism pervades all environments and operates initially to presume that I am heterosexual. Thus when I socialize outside of lesbian and/or feminist company I am often asked about a male partner; when I try to access social or health services with my lover, no spaces exist on the paperwork for us; when we walk into restaurants or stores arm in arm, people often feel compelled to ask if we are sisters or if she is my mother, although we do not look at all alike. Environments of heterosexism permit women certain identities, but deny them others: women together can be mothers, daughters, sisters, roommates, friends – but not lovers. They cannot be lovers because this is profoundly political. It is profoundly political because it both resists and threatens the oppressive system of male dominance which requires women to be heterosexual for its very existence. Compulsory heterosexuality ensures that each member of the oppressed group – women – is individually coupled with a member of the dominant group – men. This assures male rights of access to women on an economic, emotional and physical level. Lesbian existence attacks this right. It challenges heterosexual hegemony.

Thus, when I walk down the street, day or night, outwardly expressing myself as a dyke, announcing my identity as a woman who other women may have access to, but men may not, I often experience open hostility and/or violence from both heterosexual men and heterosexual women. This is intensified if I am with a group of dykes, if we are taking up space on our own terms. If women refuse/resist the heterosexual identity that is the only one available to them in heterosexist society, they will be denied any others.

It is not as simple as 'being out'. Heterosexism works either to deny me/make me invisible, or to force on me identities which, (a) do not threaten the system of compulsory heterosexuality, and (b) fit into heterosexist ideology. A sampling of the most popular stereotypes about lesbians illustrates this point: all I need is a good man; I can't get a man; I am like a man; I was sexually traumatized by a man; I want to be a man; I am a man-hater. There is an obvious common denominator here – men. Yet what defines my existence as a lesbian is loving other women: men are totally irrelevant to this basic definition. All the myths and stereotypes are related to men because, in a heterosexist society, women together are not allowed the self-defined identity. Compulsory heterosexuality demands that women direct their energies towards men, and be accessible to men. Women who direct all energies to other women cannot be accessed by men. But men must be in the picture and one way to do this is through myths and stereotypes.

While lesbian existence is very different from heterosexual existence, it would be a mistake for heterosexuals to always make it 'other'. When you think of sexuality, don't automatically think lesbian, think self. Heterosexuality is as much a social construction as lesbian sexuality is. However, as Valentine (1993c: 396) notes, 'such is the strength of the assumption of the "naturalness" of heterosexual hegemony, that most people are oblivious to the way that it operates as a process of power relations in all spaces'. Thus, for example, talking about the weekend with heterosexual colleagues, I might explain that on Friday I went to a lesbian dance, Saturday I relaxed with a great new lesbian novel and Sunday I went to see a lesbian movie. This is often interpreted as 'flaunting' or obsessing on my sexuality. My colleague can tell me that Friday night she went to a [————] nightclub with her husband, they spent Saturday morning in bed reading the [————] newspaper, and on Sunday they took the kids to a matinee to see a [————] movie. The point is that it is I who will fill the empty spaces with 'heterosexual' not she; that her (hetero)sexuality is always upfront and centre – her wedding band announces it, walking down the street hand in hand with her husband announces it, her discussions of everyday life announce it – is not recognized. The spaces above are the spaces of privilege.

WHAT'S WRONG WITH THIS PICTURE? MISSING SISTERS IN GEOGRAPHY

The silencing and exclusion of disabled women and lesbians is, not surprisingly, manifest in geographic literature. In this section, we comment on the phenomena of missing sisters in our discipline.

Invisible sisters: disabled women

From the perspective of a disabled woman, the geographic literature is in many ways a wasteland. Few studies speak to the lives of the disabled, and even fewer grapple with the social processes through which mental and physical disabilities become bases for discrimination, marginalization and oppression. A computerized search of 5,000 geographic journals found no references at all to disabled women. Issues such as access for the disabled were addressed in journals of related disciplines, in particular planning and engineering, but often from a physical rather than social planning perspective. A few geographers have studied facets of the lives of the disabled, such as service provision and coping amongst the psychiatrically disabled (e.g. Dear 1981; Elliott 1992; Taylor 1989), work on the visually impaired (Golledge 1993), and the US disabled persons' movement (Dorn 1994) and Dear and Wolch's (1987) important work on homelessness. Some geographers report that there is significant ongoing research on such topics as multiple sclerosis but, due to lack of interest, this work has not been very visible in geographic journals (personal communications). More encouraging, graduate research on disability issues is increasing as I discovered while trying to organize a special journal issue on geography and disability research.

It is important to recognize that some work by non-geographers speaks to geographic aspects of disability and oppression. For example, Hahn (1986, 1989), a political scientist, has discussed the challenges of creating more inclusive urban built environments and the role of ableist 'body images' in marginalizing the disabled.

Despite these encouraging signs, in radical geography disabled women have been rendered almost completely invisible and silent. Browse through the index of a major collection (like Peet and Thrift's *New Models*, 1989) and try to find a reference to the disabled or to ableism (I could not). Or turn to discussions of ways forward in feminist geography (e.g. Bowlby *et al.* 1989) and note major research priorities. Sexuality is there. So too is 'race, class and gender'. Then why not ableism or ageism? We know from women's lives that these things matter. Why aren't we saying so? More importantly, why aren't we doing something about it?

To be fair, the 'invisibility' of disabled women and men in radical geography undoubtedly reflects a real absence in academic and student ranks (in fact, geographers could use a good study of this). After all, performance and evaluation standards in academia are extremely ableist; allowing little or no room for differences in abilities to read, write, teach or do research, take on speaking engagements and, most importantly, 'produce' in general. Those who cannot perform to ableist standards are likely to find themselves pressured to give up their positions, even though they may be able to make important contributions, and despite laws requiring accommodation of the disabled in the workplace. If they resist, they face an often lonely battle to convey the need for non-ableist standards and practices. To modify an old adage, it is hard to understand what it means to live as a disabled person in an ableist society until you have walked (or wheeled) some miles in her shoes.

It is very likely, therefore, that many disabled women and men are quickly 'pushed out' of the system. As women are under-represented in positions of power within the discipline, it is likely that disabled women are especially vulnerable to such pressures. They lack, for instance, the chance to be mentored by women, especially disabled female faculty, and mentoring is essential to coping with the gendered power relations of the 'old boys' network'. Class oppression also comes into play, as disabled women are especially vulnerable to poverty (National Council of Welfare 1990). For disabled women, limited economic means translates into concrete difficulties in acquiring mobility aids needed to negotiate campuses as well as in raising the tuition needed to pursue higher education.

In a way, it is puzzling that radical geographers have had very little to say about these processes, especially considering the Marxist origins of this part of the discipline. For the exclusion and marginalization of the disabled is deeply intertwined with the 'commodification' of human life; with valuing people for their capacity to produce commodities, services and profit rather than for diverse talents, abilities and ways of being and becoming. This is one of the more damaging and insidious facets of patriarchal, capitalist societies for it encourages us to reduce human worth to 'what we can get out of each other' and in the process helps marginalize those who, for various reasons, cannot 'compete'. There is abundant evidence of this economic and social devaluation. In Ontario, Canada, for instance, 80 per cent of disabled persons live in poverty, a result of discrimination and exclusion in the job market, and relatively meagre public and private support programmes (Disabled Persons for Employment Equity 1992).

Of course, the silences in the literature are just one sign of academic practices which marginalize the disabled. Traditional research methods, which construct the disabled as an 'object' population rather than as experts in living as part of the disabled community, are another form of silencing and exclusion. So too are conferences which scatter sessions between buildings and floors, forcing mobility-impaired people to cover long distances and endure fatigue and often pain. In fact, at most conferences there is no sign of any accommodations for the disabled: no special information booth, no questions about special needs on registration forms, no aides available to assist people – with a wheelchair, or knowledge of sign language, or just indicating accessible elevators. Sadly, much the same can be said of our campuses, where many administrations refuse to spend the money needed to make access a reality.

Invisible sisters: lesbians

The geographic literature cannot be fairly called a wasteland when it comes to lesbians; however, analyses of the spatial expression of sexuality, have appeared only very recently in the margins of geographical research. Feminist geography is just beginning to recognize sexuality as an important part of that great abyss of 'otherness' (England 1994; McDowell 1993a, 1993b; Peake 1993). The emerging literature on lesbian and gay geographies to date, speaks much more to the experiences of gay men than to those of lesbians (Valentine 1992, 1993a, 1993b, 1993c is an exception). Nonetheless, it is very exciting to see a growing body of work on the impact of lesbians and gay men in the socio-spatial restructuring of the city (Adler and Brenner 1991; Bell 1991; Bell *et al.* 1994; Knopp 1987, 1990a, 1990b; Lauria and Knopp 1985; Winchester and White 1988). However, there is an urgent need for a much more critical approach to the research issues addressed in this literature.

The dangers inherent in discussing lesbians and gay men in the same breath should be clear given the wealth of feminist work in geography illustrating the critical difference that gender makes (McDowell 1993a, 1993b; Pratt 1993; Rose 1993). That it is impossible to ignore the fact that human experience is gendered is surely well established. It is clear, for example, that to discuss the 'working class' is to ignore (amongst other things) the all-important differences between what it is to be the 'woman on the street' as opposed to the 'man on the street'. This applies equally in the realm of sexuality. To state the obvious: lesbians are women, gay men are men and thus common experiences cannot be presumed. Although lesbians are not completely ignored, many authors – rather than recognizing that the socio-spatial experiences of lesbians and gay men may in fact constitute two separate and discrete research problems – are at great pains to explain the bases for the differences observed.

For example, Johnston (1979; cited in Lauria and Knopp 1985) argues that these differences reflect the fact that gay men may perceive a greater need for territory. Discussing this and other explanations, Peake (1993: 425) argues that:

> Such empirical and conceptual generalizations smack of an inability to rise above the level of the patriarchal mire, of being unable to unpack the heterogeneity of class, 'race', and other relations that characterize the lesbian community.

I would agree, but go further to suggest that the question of lesbians does not always have to be addressed in research focused on gay men. The obligation to attempt to explain differences in the socio-spatial experiences of lesbians and gay men only arises if the premise is that there should be any similarity. Both lesbians and gay men engage in same-sex relationships and experience oppressive marginalization, but there is no reason to assume they must have any more in common than that.

Recognizing that the question of difference/similarity need not always be a question may avoid such dangerously misleading arguments as those made by Lauria and Knopp (1985) in discussing the reasons for differences in the impact lesbians and gays have had on the city's socio-spatial structure. Part of their explanation is based on their belief that lesbian sexuality has always been more accepted than gay male sexuality. They illustrate by stating, 'lesbian sexuality has been accepted under certain conditions, as when it is "performed" for men by women who conform to societal standards of beauty' (1985: 158). They fail to realize that this is not 'lesbian sexuality'.[1] The most superficial exploration of pornography will show that the scenario they describe has nothing to do with 'lesbian sexuality' and everything to do with two or more women preparing each other for 'the main act' (i.e. heterosexual intercourse). And 'the main act' always ensures that there is no threat whatsoever to heterosexual hegemony.

Thus, given the critical importance of gender in structuring our experiences of everyday spaces, commonalities cannot be presumed between lesbians and gay men. The frequent use of the terms 'gay' and 'homosexual' to mean all lesbians and gay men both affects and denies the realities of lesbian existence. It makes invisible the huge power differential based on gender between lesbians and gay men (and doesn't even begin to touch on the differences within each community). It fails to reflect the myriad ways in which gay men, as men, oppress lesbians, as women. And, as women's experiences are more often than not subsumed under the 'norm' of men's in androcentric geographical research, so too are lesbians' experiences often subsumed under gay men's.

This problem of ignoring and/or trivializing the importance of differences based on gender – not to mention ability, class, 'race', culture and so on – points to another dangerous path that 'lesbian and gay geographies' could easily follow: one which sees a dichotomy with heterosexuality as dominant and all other sexuality thrown together into one big oppositional construct. All of us who fall into this latter category are then defined by the fact that we do not engage exclusively in 'normal' heterosexual sex. Those familiar with the politics of lesbian and of gay communities in many North American cities will be aware of the recent trend toward 'queering' everyone who is not heterosexual. For example, it seems of late that Lesbian and Gay Pride Day has become Lesbian and Gay and Bisexual and Transsexual and Transgendered Peoples' Pride Day. While this alliance may be politically expedient, it muddies the distinctions between these groups and is likely to depoliticize lesbian existence in the process. For example, the unique threat which lesbian existence poses to heteropatriarchy gets lost:[2] in this crowd of lesbians, gay men, bisexuals, transsexuals, transgendered people and straight supporters, only one group denies men access to women – lesbians. Further, as lesbian autonomy is made invisible so too is the revolutionary message to other women that there can be life without men. Many feminist lesbians, especially those who came out through the so-called 'second wave' of the modern women's movement, are wondering what ever happened to the consciously political struggle for a collective resistance/challenge to heteropatriarchy, and to our dream of the Lesbian Nation (Johnston 1973).

Blurred distinctions pose another problem: just as feminists in Anglo-American geography have painfully realized that there is no single category 'Woman' (McDowell 1993a), it must be recognized that there is no single category 'Lesbian', never mind 'Queer'.[3] Discussing lesbian organizing in Toronto, Ross points out that:

> In large urban centers across Canada and other Western countries, the 1980s have heralded the subdivision of activist lesbians into specialized groupings: lesbians of color, Jewish lesbians, working class lesbians, leather dykes, lesbians against sado-masochism, older lesbians, lesbian youth, disabled lesbians and so on.
>
> (Ross 1990: 88)

As a radical feminist Dyke who experiences the privileges of being white, formally educated and able-bodied, my resistance to and experiences of heteropatriarchal oppression, are as different from those of 'homosexual ladies to really watch out for' (Bechdel 1994), as they are from S/M lesbians. And they are worlds apart from those of gay men, bisexuals, transsexuals and transgendered people.

Part of this conflation of very different experiences stems from approaching this topic as a question of sex – of something supposed to take place between individuals in private space – rather than as a question of power and oppression. For example, in discussing negative reactions to the appearance of Knopp's (1990a) article in the *Geographical Magazine*, Bell (1991: 327–8) states:

> members of the academy do feel uneasy researching this topic. This *squeamishness* regarding sexual issues is partly homophobic and partly a 'justifiable fear of never being cited, except in a list of interesting, albeit peripheral work' (Christopherson 1989, 88); while the study of the geographies of homosexuality remains marginalized and obscure, it will not attract career- and status-minded academics.
>
> (my emphasis)

This quote raises two sets of issues. First, 'homophobia' is a very small part of the explanation of why this subject is often treated with a certain squeamishness. 'Homophobia' is to heterosexism what 'prejudice' is to racism: neither comes close to describing the systems of oppression from which those who are 'homophobic' and/or 'prejudiced' greatly benefit. Heterosexism – which privileges heterosexuality as the only true, pure and natural sexuality and discredits and makes deviant any other expression of sexuality for women – is necessary to heteropatriarchy. Heterosexism works effectively to control women, all women, and should not be conceptualized as simply irrational fear of 'different lifestyles'. It ensures that women have little choice other than to enter into intimate relationships with members of the very group that oppresses them – men. Heterosexism bestows a whole array of privileges on heterosexuals whilst encouraging hatred of lesbians through harassment, discrimination and violence.[4] In other words, squeamishness about sex simply does not come close to explaining why this subject has rarely been subjected to a critical analysis in geography.

Second, the 'justifiable fear of never being cited', can be described as part of the power relations at work in geographical research. Deciding to 'play it safe' for the sake of career and status is thoroughly understandable in a patriarchal institution; however it has very real implications for those of us who are marginalized within the discipline. As McDowell (1992b: 59) argues, 'we cannot ignore our own positions as part of the conventional structures of power within the academy, nor, although it is often painful, can we afford to ignore the structures of power between women'. Feminists doing work in this area must remain vigilant so that the question of power and oppression – who loses, who gains, in whose interest is our oppression – is not lost in efforts to make lesbians visible in geography.[5]

CONFRONTING ABLEISM AND HETEROSEXISM IN GEOGRAPHY

We want to underscore, again, the point that manifestations of these oppressions in workplaces, homes and communities often make the difference between disabled women and lesbians who can be active political advocates of change, and those who are too exhausted and discouraged to even pick up the phone. It is sobering, for example, to read that 50 per cent of all rheumatoid arthritis patients suffer from clinically defined depression. Pain and mobility limitations contribute to this, but are probably less significant than underfunding of medical research and services, and unsupportive medical practices based on some physicians' aversion to treating incurable and chronic diseases because they are more troublesome to deal with. For lesbians the penalties for coming out in the workplace and/or wider community (such as job loss and violence) are so severe that many women live a splintered existence: out in their 'private' lives, but trying to 'pass' as heterosexual in other facets of life. The damage involved in denying one's identity and culture in order to survive should not be underestimated. The personal costs are devastating; so, too, are the costs to all of us as members of society: the knowledges and experiences of disabled women and lesbians, and their capacities to understand, care and contribute become lost to teaching, research, planning and policy-making. Our stories are hidden, our voices silenced; a lost heritage for us, our children and their children.

The social construction of disabled women and lesbians as oppressed others is a pervasive and complex process; it permeates many facets of academic life and so much of daily life that, for women like ourselves, it is inescapable. In this section we consider ways of challenging ableism and heterosexism in geography, an endeavour that requires fundamental rethinking of our social theories, research methods and politics as academics.

Are geographers up to it?
Facing up to ableism

As ableism is a pervasive set of social relations, practices and ideas affecting both our discipline and society, it follows that to attack ableism within the social sciences is to touch but the tip of the iceberg. These efforts are very important in their own right, but are unlikely to challenge ableism unless they are part of a comprehensive offensive against attitudes, practices, services, policies and power relations that imprison and maim disabled people because they are 'different'. This robs society of precious sources of knowledge and hope.

If ableism is so much a part of radical geography and daily life, what can we do to challenge it? Must disabled women remain excluded and silenced 'others'?

One important first step is to insist that research by and for – as opposed to 'on' – the disabled deserves greater priority in radical geography. We must also agree that it is important to fight for inclusion of more disabled people in the radical research community, not only by trying to increase representation within academia, but with a whole array of measures ensuring the disabled physical, economic and social access to workplaces and communities. Without solid support for rethinking how we work and live, for finding more inclusionary ways of producing and living, all the affirmative action initiatives in the world won't allow disabled women to define their own revolutionary terms and conditions of participating in the production of knowledge and daily life in general. Nor will it support disabled women's struggles against sexist ableism in its various guises – including efforts to reduce disabled women's comparatively greater vulnerability to male violence (Masuda and Ridington 1992).[6]

Thinking through strategies for challenging ableism, it quickly becomes apparent that even the initial steps require us, in a very profound way, to re-learn how we value differences in ourselves and in others. Performance standards which value quantitative output (such as grants and papers) and frequent conference travel, for example, not only devalue 'ordinary' academic activities such as teaching a class, they fail to recognize just how amazing it may be for someone who is mentally or physically disabled to produce even a single paper. Somehow, we need to learn how to think and act through 'other eyes' if we hope to challenge these oppressions and exclusions. This means, amongst other things, learning to 'see', through our theories, research and lives, the relations and processes through which disabled women are socially constructed as marginal and excluded others. It means, as well, learning to respect the pain and anger of the disabled at being cast as less important and less capable; it means supporting struggles to challenge the processes of exclusion and marginalization which sustain this 'other' status.

A related task is to search for creative ways of giving voice to lived experiences of ableist relations and practices in our research designs. This means, among other things, relinquishing privileged academic viewpoints in favour of more inclusive modes of description and analysis, not simply giving 'voice' and validity to 'subjugated' knowledges (although this is important) but also developing research designs in which participants have a say in the conduct, interpretation and use of research, and where both researcher and participants 'live the research process' in a very direct way (Chouinard 1994).

It also means struggling to conceptualize disablement processes as part of the political economy of patriarchal, capitalist societies – treating disablement not just as an unrelated set of oppressive relations and practices added on to existing research agendas, but as processes rooted in significant ways in classist, sexist, heterosexist, ageist (as well as ableist) relations and practices which are part and parcel of the development of capitalist societies today. The growing body of radical and/or feminist literature can help geographers to conceptualize disablement processes in this comprehensive way. Work on the political economy of disability (Oliver 1990) and feminist critiques of the social construction of the disabled and of disabled women in both the women's and disability rights movements (e.g. Findlay and Randall 1988; Morris 1991) has highlighted, amongst other things, the importance of cultural images in distorting peoples' understanding of disabled women's lives and how these distortions affect disabled women's sense of their own realities and struggles. Thompson writes:

> Anger felt by women because of our disabilities is rarely accepted in women's communities, or anywhere else for that matter. Disabled or not, most of us grew up with media images depicting pathetic little 'cripple' children on various telethons or blind beggars with caps in hand ('handicap') or 'brave' war heroes limping back home where they were promptly forgotten. Such individuals' anger was never seen, and still rarely is. Instead of acknowledging the basic humanity of our often-powerful emotions, able-bodied persons tend to view us either as helpless things to be pitied or as Super-Crips, gallantly fighting to overcome insurmountable odds. Such attitudes display a bizarre two-tiered mind-set: it is horrible beyond imagination to be disabled, but disabled people with guts can, if they only try hard enough, make themselves almost 'normal.' The absurdity of such all-or-nothing images is obvious. So, too, is the damage these images do to disabled people by robbing us of our sense of reality.
>
> (cited in Morris 1991: 100)

Unless geographers manage to build on such sophisticated insights by, for example, considering the social construction of exclusionary territories, we risk perpetuating representations of the disabled as 'special cases' rather than as people living through some of the most destructive manifestations of societies driven by profit, greed, intolerance and superficial types of individual success – qualities which translate into excluding those of us who are 'different' from the spaces of the powerful and advantaged.

To further efforts to develop such theories (e.g. Oliver 1990), geographers need input and guidance from those living with disabilities and struggling to challenge the multiple discriminations that go along with this type of 'difference'. We need to learn to open our conceptual and empirical debates to those living disablement. This means including disabled activists in the social construction of academic knowledges about ableism and in debates about its connections to broader lived relations such as class, and letting them bring their lived experiences of discrimination and struggle into the research process. In this way, we can build political challenges to ableism, within and outside of the research process, including alliances between researchers and disabled activists.

These reconstructions, focusing as they do on 'enpowering' the oppressed, will not be easy. Indeed, as McDowell (1992a) points out in her discussion of feminist methods, even the most progressive research designs raise very difficult ethical, practical and political questions. The researcher is never really 'outside' the dilemmas of radical research but constantly struggles to handle them a little bit better, a little more fully.

Sensitivity to issues like exploitation of the 'researched' is likely to be especially important in the case of geographic research focusing on disabled women. These women are (at least) 'doubly disadvantaged' by gender and by mental or physical challenges. In most cases, limited finances and marginalization combine to limit their capacity to participate in society and in the research community. This means two things. First, researchers have a responsibility to further struggles to open the research process to disabled women and make their voices heard in the conduct and use of research. Second, politically, the research must be aligned with struggles against ableism and the relations, institutions and practices that support it. The method will vary from project to project, however it should be recognized that a research project which does not centrally contribute to the research and political priorities of the disabled women involved is exploitative and oppressive.

Challenging ableism in geographic research will be as revolutionary for our understanding of processes of urban and regional change as it will be for the politics and practices of research. For disabled women, it will help us better understand how exclusion, silencing and oppression are reproduced within the predominantly white, middle-class ranks of the women's movement, as well as through patriarchal institutions. It will make it as important to understand the positioning of people within the 'micro-relations' of power in daily life and life spaces, as it is to understand the role of major social divisions in empowering some groups at the expense of vulnerable 'others'. More importantly, challenging ableism will force us to grapple with time- and place-specific manifestations of ableism, and with how the living of these oppressive and exclusionary relations translates into resistance and rebellion.

Are geographers up to it? Facing up to heterosexism

Much of the geographic work to date on lesbians and on gays has concentrated on increasing visibility rather than critical analyses of heterosexual hegemony. That is, most authors concentrate on lesbian space and gay male space (e.g. meeting places, gentrified neighbourhoods) rather than on everyday spaces – the heterosexual and hostile environments in which lesbians and gay men spend most of their time. Of course, given the dominance of heterosexuality in space, it is important to document and understand lesbian space and gay male space, but if this is all we do then it is surely a case of adding 'queers' and stirring. It is striking how little critical political analysis there is in the literature, and how few connections are made to the wealth of feminist work in geography. Valentine's work is an important and refreshing exception in that it moves beyond the 'impact on the city' approach to a critical examination of how lesbians create, transform and negotiate not only lesbian environments but the more day-to-day environments of heterosexism (see especially Valentine 1993b, 1993c). She argues (1993a: 114) that more research is needed 'to gain a better understanding of how heterosexual hegemony which is so often taken for granted, is reproduced in space'.

As public discourse in North America on lesbians changes, it is more critical than ever to illustrate the ways in which our material realities have not; that is, the ways in which heterosexual hegemony prevails. As Westenhoefer wryly points out in the infamous *Newsweek* article that put lesbians on almost every news-stand across North America, 'We're like the Evian water of the '90's. Everybody wants to know a lesbian or to be with a lesbian or just to dress like one' (21 June 1993). Knowing a lesbian may indeed be 'trendy' in parts of white Western culture but does this manifest itself in political solidarity? Does it lead to thousands of heterosexuals who 'know a lesbian' marching for lesbian rights? Does it lead to these same heterosexuals insisting on placing lesbian literature in the school system? Does it lead to mass heterosexual mobilization around the heterosexual bias of immigration, taxation and adoption laws? A change in discourse alone will not challenge structures of oppression.

Further, a change in discourse is often a double-edged sword. Although changing discourse undoubtedly has political implications, it can also make lesbians' struggles against heterosexism more difficult, as the ideology of heteropatriarchy works to suggest that we have less to struggle against. An analogy can be easily drawn: consider the feminist struggle against male violence against women; everyone is talking about it, it is squarely on the public agenda, but women are no less likely to experience male violence today than they were twenty years ago (Bart and Moran 1993; Dobash and Dobash 1992; Statistics Canada 1993; Walker 1990). Yet, despite ample hard evidence that this violence against women is endemic to heteropatriarchal culture, we are still likely to hear (and prefer to believe), that this violence is a product of poor anger management, women's low self-esteem, dysfunctional families and/or learned behaviour. Similarly, everybody may be talking about (even to) lesbians in the 'gay nineties', but heterosexism is still as powerful as, if not more powerful than, it was twenty years ago. Popular culture may tell us that being lesbian is fairly acceptable today, but there is no evidence that lesbian teachers are coming out in their thousands, that lesbians are holding hands in the street, that federal laws have been changed or that violence against lesbians is decreasing. It is important to make lesbians visible in geography, but the concentration on lesbian and/or gay space has so far come at the expense of critical analyses of environments of heterosexism. It is time for more of the latter, and less of the former.

REINVENTING OURSELVES AND 'THE PROJECT': MISSION IMPOSSIBLE?

The kind of geographic research envisioned in this chapter has the potential to revolutionize our understanding of social movements and struggles.[7] By learning to understand ableism and heterosexism as embedded in other social relations of power in patriarchal capitalism, and by exploring the diverse ways in which our experiences of oppression are socially constructed (reflecting our diverse locations within power relations and the role of place in social processes), geographers will be better able to explore how differences and lived identities help shape progressive political action over time and space. Another benefit, as we have learned in the course of collaboration, is that researchers can confront and challenge their own complicity in the construction of relations, identities and practices which marginalize 'other' sisters. The challenge is to do nothing less than reinvent our ways of being involved in research and other social processes, and to develop a progressive, collective 'Project' inclusive enough to ensure that 'missing sisters' are put back into the geographic picture. This does not mean each and every research project will be directly concerned with ableism or heterosexism, but it does mean that researchers will deliberately identify the 'locations' from which they are speaking, and the fact that their findings may not speak to the experiences or struggles of 'missing sisters'. It also, of course, means efforts to include the lives of disabled and lesbian women as significant facets of women's lived identities, histories and geographies (e.g. Bondi 1992; Chouinard 1992, 1994; Deutsche 1991; Duncan and Goodwin 1988; Fraser 1989; Harvey 1989; hooks 1990; McDowell 1992b; Mouffe 1988; Rooney and Israel 1985; Soja 1989).

In what follows, we consider the possibilities for such a reconstructed progressive 'Project' in radical geography (again, including feminist geography). We argue that geographers are well placed to face this challenge but that we must recognize significant barriers to a successful 'redrawing' of the 'Project'.

Mission Impossible?
Non-ableist geographies

Both the geographic and non-geographic literature show growing interest in the uneven development of peoples' capacities to contest social oppressions (e.g. Chouinard 1989; Duncan and Goodwin 1988; Fincher 1991; Ley and Mills 1993; Miliband 1991; Murgatroyd *et al.* 1985). In geography, place, identity and culture have been increasingly important themes (witness, for example, special sessions at the 1992 Institute of British Geographers Conference and the 1994 annual meeting of the Association of American Geographers). The time is ripe, then, for geographic interpretations of the complex, diverse and often fragmented processes of the social construction of lived identities, and of related processes of 'disempowerment' and marginalization.

Geographers, like social researchers in general, have robust critical traditions of enquiry on which to build reconstructed non-ableist and non-heterosexist research. Almost three decades of radical research, enriched by feminist work and by relatively new interests in postmodernism and inter-pretative theory, have provided a sound foundation for exploring processes of identity formation, empowerment and oppression, and the significance of place in those processes (e.g. Pratt and Hanson 1994; Knox 1991; Ley and Mills 1993; Peet and Thrift 1989; Rose 1993).

If there are grounds for optimism, there are also grounds for cautious concern about the possibilities for non-ableist and non-heterosexist geographies.

One challenge is the problem of having to confront our own attitudes and practices. Do you turn away when you see a severely disabled young person in a wheelchair – or do you look, watch how they cope with the environment and think about how you and others could make their lives easier by altering that environment? Do you think of them as disabled first, and a person second – do you help socially construct their identity as someone to be pitied, a victim – or do you recognize the importance of breaking through your images and reactions to build a more inclusive 'Project' in geography? Even as a disabled feminist researcher, I sometimes have difficulty putting the commonplace reactions and stereotypes aside: I am as much a product of the ableism that pervades our societies as anyone else. The point is that we must face up to our complicity in ableism if we are to challenge it effectively in our research, teaching and daily lives.

Another challenge involves treating confrontations with ableism (and heterosexism) not simply as a new topic or 'oppressed group' to study, but as a catalyst for rethinking geographic processes. For instance, in writing this chapter we have both learned that understanding of space and territory is revolutionized by understanding that ableism, heterosexism, and ageism are in many ways as significant bases of social oppression as gender, class or race. Both disabled women and lesbians are engaged in struggles for spaces of resistance and access to the territories of the powerful, fighting to gain control over the terms and conditions of their lives. That fight includes struggles over the means of producing physical and social environments, such as cultural media, schools, popular stereotypes, capital and state programmes.

Related challenges include struggles for the presence and visibility of the disabled in social spaces. Some catalogues, for instance, have started to include disabled models – a small step, but one that reminds us of the presence of disabled persons in our communities. Sadly, other signs point to the continued exclusion and marginalization of disabled people. For example, in Toronto a disability network information programme once broadcast at noon on Saturdays is now shown at 6 am. The sociocultural message is clear: the concerns of disabled people are not sufficiently important to warrant broadcast at a time when people are likely to be awake.

A non-ableist geography will not only have to be sensitive to complex processes of access and exclusion, but also to how disabled people must struggle to create liveable, supportive spaces. It will also require research processes which include disabled persons as vital contributors; this in turn will require meaningful political alliances with disability activists and organizations. In short, it will require nothing less than a redrawing of 'The Project' to include the experiences and struggles of the disabled as vital facets of contemporary social movements: not conceptualizing the disabled or lesbians as 'others' but as significant voices at the margins of social power; people who are as much a part of humanity as their able-bodied and/or heterosexual sisters.

Is such a reconstructed 'Project' an impossible dream? Only if we lack the geographic imagination to envision reinventing ourselves as researchers and teachers determined to fight complicity in the social construction of 'others', including relations and practices that create spaces which render our sisters invisible, silenced and oppressed. Will we find the courage to give up some of our privilege, for example by refusing to participate in events that are inaccessible to our disabled sisters?[8] Will we find the courage to say loudly, often and publicly that we will not be part of any 'Project' in which disabled women are cast as the 'other'? It will mean insisting that by including disabled women, on their own revolutionary terms, we are all made stronger, more complete, and better able to understand and fight our oppressions.

Mission impossible? Non-heterosexist geographies

A feminist critique of compulsory heterosexuality is long, long overdue in geography. The central tenets of feminist geography make it most appropriately placed to tackle heterosexism, but it will remain intellectually and politically impoverished if it ignores the fact that patriarchy is hetero-patriarchy. How are we, as feminist geographers, to expose and challenge the patriarchal structures that maintain male supremacy without recognizing the critical role that compulsory heterosexuality plays in the social control of women? Both lesbians and straight women have an interest in challenging heterosexism. As Valentine (1993c: 411) points out:

> by ignoring antilesbianism or collaborating in perpetuating it, some heterosexual women comply in their own oppression, because such antilesbianism is also used to police heterosexual women's dress, behavior, and activities. Hence, if 'dyke' were not a term of oppression,

heterosexual women would also have more freedom to define their own identities.

This is not a difficult concept to grasp and I challenge straight feminist geographers to recognize their own complicity in the construction of dominant and oppositional sexualities, and thus in the oppression of all women. Recognize heterosexual privilege and critically analyse the system of rewards that is based on heterosexuality. Understand how this system works to construct and reinforce that identity; how it helps maintain compulsory heterosexuality, without which heteropatriarchy could not exist. This involves incorporating a recognition of heterosexism in both theory and practice.

If our work is truly to be part of the struggle for social change, the struggle for all women's liberation, then a challenging discourse in geography is not enough. The connections between the production of knowledge and the politics and practices of research are obvious. The next time someone calls you (or a woman you know) a dyke, or hints that you are, look at why this has happened. How have you stepped out of line, how have you moved beyond the heteropatriarchal definition of a 'woman'? More critically, look at your response: do you quickly mention your male partner, apologize for that behaviour, try to justify what you are doing, and/or step back into line? Or do you challenge, disrupt and unsettle? Do you recognize lesbian-baiting for what it is and do you recognize the power of heterosexism to control women? Do you challenge it, and say, 'Why, thank you'? We need discourse, but we also need politics and practice.

It is impossible to think of ways to move towards a non-heterosexist geography without raising some questions about representation. The spatial analysis of sexuality may well be of interest to many feminist geographers, but it would be a mistake for them to presume that this necessitates an investigation of lesbian activity, community and/or space. The structures of heterosexuality are crying out for a critical analysis by heterosexual feminist geographers; this would be a fruitful and appropriate area of study, and one which may avoid appropriating the voices of 'others'. This question of who should do what type of research is dealt with in a challenging collection of papers in *The Professional Geographer* (England 1994; Gilbert 1994; Katz 1994; Kobayashi 1994; Nast 1994; Staeheli and Lawson 1994). Staeheli and Lawson are well worth quoting at length:

> The most notable difficulty expressed in these papers is the assumption that feminist women researchers have insider status based on their experiences and training that makes women 'sisters' with the women they research. In the field and in the community, however, it is becoming clear that this assumption is naive, and perhaps dangerous. The assumption of insider status ignores the various dimensions of difference that distinguish women and the issues with which they are concerned. This realization is generating a sense of crisis as feminist researchers question their relationships with the people, places, and power relations they study. In its most extreme form, it can lead to guilt, paralysis, and abandoning research projects.

(1994: 97)

This 'crisis' can create an atmosphere where the political runs the danger of getting too personal. I want to present an honest plea for the making of enough space for feminist geographers to discuss the very real differences amongst us. We must have the courage to recognize our own complicity in others' oppression. For example, writing together as a Dyke and a disabled woman, we, the authors, both recognize the privileges we gain from each other's oppression (see also note 6). We also understand that, although common threads run through our oppressions, our experiences often differ and neither of us would ever imagine she could speak for the other. I am not necessarily arguing that heterosexual women should never discuss lesbians in their research. Neither am I arguing that it is always okay to do so. It is important to listen to the voices of 'marginalized groups' when they talk of who should and should not do research with them. Surely feminists who are sympathetic to the concerns of, for example, lesbians, are also sympathetic to their arguments, and concerns about representation and appropriation?

This issue could make even the most ethically and politically aware feminists hold on to our academic privilege of choosing a research area. Although it may be appropriate for lesbians and straight women to work together in some cases, it is not always so. The challenging question is: do we, as academics, still retain the right to decide what is appropriate and what is not? An integral component of power and privilege is being able to make up the rules.

Further, in unearthing the failings of androcentric research on women, feminists have surely learned that marginalized groups do not speak to their oppressors in the same language that they use with each other. As we consider ways to avoid perpetuating heterosexist geographies, it is important to know, for example, that many lesbians, myself included, have no interest in being involved in any research on lesbians conducted by a straight woman. This does not mean she is a 'bad' woman; just that she will always see my experiences through 'straight' eyes. It is important for lesbians to present their own experiences. In a similar vein, Kobayshi (1994: 74) argues that:

> Political ends will be achieved only when representation is organized so that those previously disempowered are given voice. In other words, it matters that women of color speak for and with women of color.

CONCLUSIONS: TOWARD EMPOWERING GEOGRAPHIES

In this chapter we have argued for reconstructing 'Projects' in radical geography to engage with the perspectives developed from positions of exclusion and marginalization within the discipline. For the two of us, writing this chapter has enriched and challenged our geographic imaginations. We have moved from 'Projects' which fail, to questioning why and how there are 'missing sisters' in geography, toward 'Projects' which constantly challenge the taken-for-granted nature of 'otherness'. We have moved from a politics of exclusion toward a revolutionary politics of inclusion. We have used examples from the positions we know best, but our general arguments have relevance for other 'others', including radicals with perspectives that continue to unsettle the establishment. We challenge readers to take this journey; to openly confront the white, able-bodied, heterosexual, male paradigms which continue to dominate the discipline, and rediscover the diversity of lived geographies of oppression and struggle. There will be mistakes and setbacks along the way but the path is rich with possibilities for discovery and transformation.

Among the issues we have taken from this consideration of our own positions is that challenging the silences and absences in our own ranks is the first vital step toward developing less elitist knowledges and political practices. Our intention in recounting some of our own personal/political experiences is not to evoke sympathy nor apologies – when we, as two white women, act in racist ways, apologies may make us feel better but they do little to challenge the structures of oppression. Action is what is required.

Another conclusion to be drawn from this chapter is that our understanding of the uneven development of late patriarchal, capitalist societies will be woefully inadequate as long as our disabled and lesbian sisters continue to be missing. Not only will we miss diverse lived relations of oppression and resistance, but we will also impose a false homogeneity on manifestations of oppression and exploitation in peoples' lives and environments. This kind of conception of social experience and change, like Marx's 'commodity fetishism', helps blind us to the very complicated ways in which people's lives are entangled within, and altered by, processes such as the economic and social devaluation of particular labourers on the bases of ability or sexuality. The point is not simply that we need to be sensitive to particularity. The point is, rather, that such processes are central to the uneven development of patriarchal, capitalist societies.

Radical geographers are well placed to take up the challenges posed in this chapter. A sensitivity to the place-specificity of processes of oppression and resistance, and to the play of differences within those places, permits sophisticated explanations of the politics of creating and recreating 'otherness', of how 'micro'-processes of power and oppression fit within the big picture of societies driven by enduring classist, racist, sexist, heterosexist, ableist and ageist power relations or structures.

This would include exploring how these relations give rise to the economic and political inequalities which help divide and silence us, often in the company of our own, and to the differences and commonalities in the marginalization and resistance of oppressed others. These explanations would also address the social consequences of lived oppressions: for example, the links between different types and levels of representation within local states and peoples' capacities to contest and change oppressive relations in particular places. Finally, such explanations would help move us closer toward theories and methods which take structure, agency, experience, discourse and interpretation equally seriously.

We believe that in attempting to reinvent 'The Project' it is important to recognize that issues of determinism, functionalism, structure and agency and so on are not exclusive to the new post-structuralist and postmodern perspectives and debates. They are issues that other radical geographers such as Marxists and feminists have engaged with for a long time in their efforts to come to grips with changing geographies of marginalization, oppression, resistance and revolution. Much Marxist work is a rich and ongoing legacy, not a process of debate and research which somehow 'froze' in the late 1970s with clashes between so-called humanist and structuralist Marxists (Anderson 1980; Thompson 1978). It will be our own loss, and a serious one, if intolerance and misrepresentation causes us to disregard the work of the 'old' and the 'new' Marxists in our efforts to refine frameworks that focus on peoples' experiences and actions as forces of societal development and change. The same danger is present in current engagements with feminist geography, as its themes and issues are more widely considered in the radical geographical literature. For it is clear that the terms of engagement are not always set by feminists: although at least a mention of feminist geography now appears to be obligatory in radical literature, it is often little more than that. As McDowell (1992b: 58) argues:

At last, it seems as if there is a growing recognition of feminism's fundamental understanding that the deployment of the universal is inherently, if paradoxically, partial and political. But . . . it also seems clear that this recognition is based not on an understanding of feminist arguments, and more certainly not on a commitment to feminism's revolutionary project, but rather on an eclectic reading of postmodernism

Unless we can link the causes of oppression and practices to their human costs in particular places, we are likely to be left with idiosyncratic and elitist interpretative accounts of social experience and action. A focus on social relations forces us to reconsider, time and again, our own positions in processes of creating and using knowledge; without that kind of 'reality check', that reminder that we are part of pervasive power structures, we are likely to become entangled in our own interpretative webs. For example, in situations where the 'deconstructionist' author is textually represented as 'in the background', giving voice to 'others' and challenging authority, while in practice relinquishing little, if any, power over the production and use of knowledges (see for example the interesting exchange between Massey 1991, Deutsche 1991 and Harvey 1992).

Radical geography will be impoverished, politically as well as intellectually, if we cannot find ways of balancing concerns with diverse experiences of oppression and exclusion, with continued struggles to advance our understanding of the causal processes helping to perpetuate sexism, heterosexism, ableism, racism and ageism. We hope this chapter has indicated that a solid foundation for projects in radical geography must draw at least as fully on existing traditions like Marxism and feminism as on proposals for more 'interpretative' theories and methods. We have to get at the causes of power and oppression, as well as noting their diverse manifestations in people's lives and locations, if we are to speak to and act on issues of social change. If we do not, not only do we impoverish diverse knowledges and help silence voices, but we also risk a political limbo in which it is quite unclear why academics should be speaking for anyone rather than stepping aside and letting people speak for themselves. We are either trying to act as facilitators of change, in part by using our own skills and privileges to identify the causes and consequences of social oppression, or we are parasites exploiting 'others' for dubious reasons.

We also hope we have indicated why the social construction of 'otherness' is an issue for all of us. To fail to listen to, and address, the voices of lesbians and disabled women, amongst other 'others', is to cut off crucial avenues of debate and research – work and connections that could benefit us all by challenging current, exclusionary research practices and engaging with 'others' not on elitist terms but on their own revolutionary agendas. It should be clear that we are not simply looking for a piece of the pie – we want to change the recipe.

ACKNOWLEDGEMENTS

The authors would like to thank Kim England, Joan Flood, Larry Knopp and Linda Peake for insightful comments on an earlier draft of this paper.

NOTES

1   In fairness, the authors' approaches to the subject have changed considerably in the last ten years (personal communication with Larry Knopp).
2   The term 'heteropatriarchy' is taken from Valentine (1993c: 396):

> To be gay [sic], therefore is not only to violate norms about sexual behavior and family structure – but also to deviate from the norms of 'natural' masculine and feminine behavior. These norms change over space and time, and hence sexuality is not merely defined by sexual acts but exists as a process of power relations. Heterosexuality in modern Western society can therefore be defined as a heteropatriarchy, that is, as a process of sociosexual power relations which reflect and reproduce male dominance.

3   Valentine (1993b) includes an interesting discussion of essentialism versus constructionism in understanding the roots of lesbianism.
4   For example, Valentine (1993c: 408) cites a San Francisco study of 400 lesbians: '84% had experienced antilesbian verbal harassment, 57% had been threatened with physical violence, and 12% had been punched, kicked or beaten . . . '
5   I want to be very clear that in using Bell's quotation from Christopherson's 'On being outside "the project"' article, and using it somewhat out of context, I am not criticizing Christopherson per se.
6   As Masuda and Ridington found in the DAWN study (1992: vii):

> Of the 245 women who participated in this survey, 40% had been raped, abused or assaulted, 64% had been verbally abused; girls with disabilities have less than an equal chance of escaping abuse than their non-disabled sisters; women with multiple disabilities experienced multiple-incidents of abuse, and only 10% of women who were abused sought help from transition houses, of which only half were accommodated.

7   For example, Vera is working on a project concerned with disabled women's struggles. Among its innovative features are an advisory group composed of

disabled women activists (to oversee and discuss the work from start to finish), and a conception of social experience and change in which efforts to contest ableism are understood in connection with other relations of oppression. These features will help ensure a research process closely guided by the expertise of disabled women activists, a nuanced explanation of the socio-spatial barriers that disabled women are contesting in a range of social movements, and recognition of the diversity in the political lives of disabled women. It will be a geographic research project by and for disabled women.

8   An example of giving up a little piece of privilege: I live in a town where is only one lesbian dance every month or two, so these dances are very important to me. However, when a dance was arranged in an inaccessible facility, I knew that, despite the fact that I *wanted* to go, I could not. By using my able-bodied privilege, I would have participated in sending out a clear message that disabled women were not welcome at a dance that advertised 'all women welcome'. As simply not going in itself would not make any kind of statement, I joined with other lesbians and stood outside the dance to let other women know why we weren't going in. This was a small matter, but it did make a difference: there has not been a dance in an inaccessible building since.

PART III

(RE)SEARCHINGS

ENGENDERING RACE RESEARCH

Unsettling the self—Other dichotomy

Kay Anderson

INTRODUCTION

At a moment in feminist theorizing when scholars are grappling with ethno-centric presumptions of a 'generic woman' implicit within 'imperial feminism' (Amos and Parmar 1984), it is timely to note the paucity of attempts to unsettle the epistemology of separation implicit in much race research. The fictionalized collectivities of 'Black', 'White', 'European', 'Asian' and so on – the stock in trade of the field called 'race relations' – are often the corollaries of a dichotomized us/them framework that (unwittingly) obscures the sub-jectivities of identities internal to those categories. Such a framework also tends to overwrite the interconnections of privileged race positions with other sources of identity and power. Whereas the critique of Western feminism by Black, post-colonial and lesbian writers has challenged feminist consensus (Butler 1990; Collins 1991; hooks 1981, 1991; Larbalestier 1991; Singleton 1989), much race research – including work by anti-colonialists such as Said (1978) and Clifford (1988) – has worked with modernist presumptions of an ordered (racialized) reality whose subject positionings are, for the most part, fixed and undifferentiated (c.f. Anthias and Yuval-Davis 1992; Donald and Rattansi 1992).

In this chapter I seek to problematize the polarity of race identities upon which rests the cohering argument of my earlier work *Vancouver's Chinatown* (1991). I aim to undertake such an auto-critique by feeding into the China-town story the discursive fields and social positionings of gender and sexuality, a task I undertake not for its own sake, but rather to sharpen the critical analysis of the many valences of social power. By extension, the chapter critiques other work in race relations that implicitly or explicitly disengages race identities from other historically situated oppressions such as those surrounding gender, class and sexuality. Without discrediting work that specifies the contribution that race-based oppression makes to structures of inequality, the chapter seeks to foreground the *multiplicity and mobility* of subject positionings, including those of race and gender.

Such a style of analysis may be particularly revealing, because while racism has long structured socio-spatial relations in British settler nations such as

Canada and Australia, it has been woven into a range of power-differentiated regimes out of which colonial relations have been organized into the present. Certainly the stories that emerge from a re-examination of select moments in the history of White/ Chinese relations in Vancouver, Canada, reveal a more 'complex dominator identity' (Plumwood 1993) than a unified White oppressor. The projects of colonialism, themselves manifestly variable from place to place, relied on the imaginative and practical leadership – less of Whites *per se* (as if there exists such an abstract, uncontradictory 'self'; see Bhabha 1990a) than – of a specific 'master subject' who was White, adult, male, heterosexual and bourgeois (Rose 1993). To erase these refinements to the dominator perspective is to risk invoking a falsely tidy dichotomy of relations between *a* racialized 'us' versus *a* racialized 'them', when in reality the social processes constituting social relations were complex and differentiated. The oppressions through which 'colonialism's cultures' (Thomas 1994) were elaborated in Canada's western province of British Columbia had myriad and overlapping sources in the structures of capitalism, patriarchy and cultural domination by race and sexuality. And while pronouncements about the intersections of diverse idioms of oppression are now commonplace in theorizing in the social sciences – with many efforts at formulating multi-dimensional models of class, race and gender oppression (see e.g. Bottomley and de Lepervanche 1984; Bottomley *et al.* 1991; Jennett and Randal 1987) – empirical demonstration of the ways in which oppressions interacted and became mutually confirming is not extensive (although see e.g. Bear 1994; Pettman 1992; Ware 1992). Within the space constraints of a chapter, this is the challenge of the first section of what follows.

Racialized and gendered discourse did not always, however, operate in a fully efficacious complicity in late nineteenth and early twentieth-century British Columbia. To re-tell the story of Vancouver's Chinatown by emphasizing the power of an (albeit more differentiated, that is, European male) centre is to risk reifying further the master-perspective of the dominator identity. Indeed at certain moments in the history of Vancouver's Chinatown, alliances between White women and Chinese men trouble the falsely consensual understanding of domination that arises from one-dimensional race analyses. Thus, if we reorient the story of *Vancouver's Chinatown* around the subjectivity of White women, at least in select moments, we shall see the spaces where are upset images of the monolithic societal racism upon which neat race narratives depend. By foregrounding such spaces, as I do in the second part of this chapter, we begin to 'denaturalize' (Kobayashi and Peake, 1994) the racialized and gendered marking of subjects that has been so central to strategies of domination (see also Jackson 1994).

A parallel oversimplification in the governing logic of 'European hegemony' is the construct of *a* homogeneous racialized category pitted beneath one coherent oppressor. Yet the gender- and class-differentiated experiences and statuses within the category 'Chinese' also defy the essentialized configurations of binary (self/Other) models. By briefly drawing on published sources documenting the uneven experiences of Chinese men and women in early Vancouver, the second part of the chapter also unsettles notions of a stably positioned, internally unified and uniformly oppressed victim. Taken together, the examples support post-structuralist critiques of the 'centred subject' (see e.g. Donaldson 1992; Nicholson 1990) by highlighting the contradictory, multi-dimensional and strategic quality of identities. The examples also go some way to demonstrating how the relationships between dominant and minoritized groups are crossed not only by diverse discursive fields (Lowe 1991), but also by multiple *positionings* that are not reducible to the binary division of 'us' and 'them'.

The potential infinity of the fractual patterns of social relations might appear to paralyse the quest for explanation conceived around a single, controlling point of determination. If we dissect categories too far, we risk losing sight of the structuring threads of power that cohere in empirically specifiable ways. That case has been persuasively argued by 'post-postmodernists' such as Walby (1992). Certainly the argument in what follows should not be taken as a refutation of the force, persistence and profound material effect of that pernicious ideological and material regime that is racism. On the contrary, to specify the intersections of axes of oppression, as this chapter undertakes to do, is not necessarily to disperse or disable the critique of power. It is rather to offer glimpses of two things. First is revealed the often mutually constitutive boundary-making practices out of which colonialism's cultures were constructed in the late nineteenth- and early twentieth-century setting of western Canada. That racism often drew on gendered meanings and positionings in these processes is not to deny racism's strength; rather it is to appreciate the wider discursive network in which racism was inserted. Second is exposed the possibility of rupture of racialized regimes (and the readings they support) when alternative speaking perspectives to that of the elite master-subject are positioned at the centre of analysis. If racialized alliances are at times crosscut by gendered and classed struggles, such as we shall see occurred in Vancouver in the 1930s, and we resist assimilating such struggles into an epistemic regime of race domination, then we glimpse fresh (Chinatown) stories and alternative political alliances and possibilities.

This chapter now turns to a brief résumé of the work which I then propose to critically revisit in the light of these introductory comments.

VANCOUVER'S CHINATOWN (1991): A BRIEF RÉSUMÉ

My intention in writing *Vancouver's Chinatown* was to reconceptualize an enclave which had long been theorized in the social sciences as a colony of the East in the West, an 'ethnic' neighbourhood whose residents and streetscapes existed in natural connection to their Oriental difference and Chineseness. In contrast to that model, I developed an anti-essentialist conceptualization of Chinatown as a construct of Western imagining and practice.

Using the example of Vancouver, Canada, where people of Chinese origin settled in a few blocks of that city's East End from the late 1870s, I traced through time the discursive practices that shaped the definition and management of that district. This, from the time of negative stereotyping between the late 1880s and 1930 when Chinatown was classified by White Vancouver society as a vice district; through the period of the 1930s and 1940s when Chinatown's classification grew more complex and contradictory; that is, when the vice classification came to coexist with the district's first formal tourist definition as Vancouver's 'Little Corner of the Far East'. During the 1950s and early 1960s – the post-war era of modernist urban planning – Chinatown was targeted by Canadian federal and civic administrations as a 'slum' and came close to being completely destroyed by urban renewal and freeway plans. Come the 1970s, Chinatown became classified as an ethnic and heritage district – valued by White Canadians precisely *for* its Chineseness and refurbished in a radically new kind of targeting as an Oriental district with funds from all three levels of Canadian government. In that project, they were joined by Chinatown retailers who, in the rush for the spoils of multiculturalism, manipulated to their own advantage the racialized representations of identity and place that delivered them to White Vancouver society.

Lying beneath the phases of neighbourhood definition, I argued, was the continuity of a racialization process that is the book's structuring narrative. Over the hundred-year period, Chinatown was constructed – both ideologically and materially – out of manifestly variable guises of race classification on the part of those armed with the conceptual and instrumental power to define and regulate the area. The role of the three levels of Canadian state in sponsoring and enforcing that power is highlighted throughout. Chinatown was not just the object of biased depiction and 'prejudice', then, as liberal theses had argued (Ward 1978), but – following Foucault (1979) – of a particular cultural politics of discursive production that enabled one (European) set of truths to acquire the status of truth and normalcy. That this operation entailed a will to dominate ('hegemony') had already been persuasively argued for a different scale and context by Said (1978). Thus, Chinatown, like that mythical region of Western imagining called the Orient, was recurrently White Vancouver's Other, I argued, a place through which a dominant group forged its own cultural understanding of its identity, boundaries, status and privilege. China-town was a site through which were articulated diverse narratives of race, health, vice, civility, blight, heritage and ethnic pluralism. Unlike other critiques of orientalism, however, notably Said and the more recent work by Lowe (1991), my interest lay not only in the discursive struggles surrounding identity and Othering strategies, but also the social production of the district, its changing material form and fortunes.

NATION-BUILDING IN COLONIAL CONTEXT: DISPLACING THE NARRATIVE DEVICE OF UNITARY RACE POSITIONINGS

In late nineteenth-century British Columbia, notions of in-group and out-group drew on a complex network of raced *and* gendered discourses. Later in this section of the chapter, we shall see the interaction of these meanings at the scale of 'place', with specific reference to the discursive construction of

Chinatown in early Vancouver. But the processes were equally evident at the scales of 'nation' and 'province'. The making of 'Canada' in its symbolic dimension entailed representational practices that were deeply saturated with race and gender concepts, and by highlighting their collusion, we further refine the identities and subjectivities out of which a dominant imagined geography of nation grew (see also Bhabba 1990a).

That Canadian officials of the late nineteenth century were seeking to create a White Canada was abundantly clear in the languages and debates recorded in the government texts that were the primary data sources for *Vancouver's Chinatown* (Anderson 1991). Federal legislation in the form of a head tax – passed in 1885 to contain Chinese immigration – thus sought to limit family settlement, while permitting a controlled amount of Chinese labour and capital. Bound up with the impulse to contain Chinese numerically was also a desire to prevent what was called 'miscegenation' or 'mixture of races'. To that end, the ultimate target of legislation appears to have been less the 'Chinese' as a collective racialized category, than the more narrow category of Chinese women. Moralities of race *and* gender fed an interactive discursive network. In the words of the Royal Commissioner for Chinese Immigration, John Chapleau, when arguing the case to the House of Commons for a head tax in 1885: 'If they came with their women they would come to settle and what with immigration and their extraordinary fecundity, would soon overrun the country' (Canada 1885b: 98). Canada's Prime Minister of the time, John MacDonald, held the same opinion: 'If wives are allowed, not a single immigrant would come without a wife, and the immorality existing to a very great extent along the Pacific Coast would be greatly aggravated' (Canada 1887: 643). Not just idle polemic, such views shaped policies that imposed constraints on Chinese family life in Canada and China well into the twentieth century (see Li 1988: Ch. 4). (By 1938, the Vancouver press reported that the 'ultimate solution' to the Chinese problem had been found in the severe sex imbalance in the local Chinese population which was constraining its ability to replenish itself, see *Province* 28 February 1938.)

Embedded in the projections of officials such as Chapleau and MacDonald was a particular construction of Chinese women, as wives and ipso facto reproductive beings who threatened the demographic strength and integrity of White Canada. Chinese women were never seen as single immigrants with the potential for waged (or unwaged) labour, yet we know that such women undertook a range of jobs within the enclave economy of Vancouver's Chinatown in the late nineteenth and early twentieth centuries (see Adilman 1984). And if Chinese women were seen by Canadian officials as anything other than 'fecund', they were cast as prostitutes – a still more ominous identity according to John Chapleau, in 1885, because 'they bring with them a most virulent form of syphilis and in a special way corrupt little boys' (Canada 1885b: xii). Such pronouncements filtered into media texts across the country to feed the image of a disease-bearing race with whom sexual liaisons would be ill-advised.

A sharpened gender awareness of the discursive processes at work in Canadian nation-building brings into view not just Chinese women immigrants, however, caught between White cultural superiority and male power. Running through the official constructions of Chinese immigration – and especially during the lead up to the Chinese Exclusion Act of 1923 (see Anderson 1991: 132–41) – was a racialized *and* gendered aesthetic that interconnected the spaces of nation and body. 'Canada' appears scripted in official texts as a pure space, one that if impregnated by the flow of alien material would become contaminated and offer up inferior 'stock'. In Figure 12.1, for example, taken from the Vancouver press during the anti-Chinese riot of 1907, we see how the iconic body of White woman was grafted onto the space of nation (and province). The rhetorical device within this system of representation was to symbolically construct the nation as passive and pure by affording it the attributes of an Anglo female body. Like other symbolic figurations of nation such as Miss Britannia and Lady Liberty (see Pateman 1988; Yuval-Davis 1991), Miss BC is made to emblematically stand as an essence under peril of violation. Thus invested with agency is the wilful guardian of nation, Prime Minister Wilfrid Laurier, whose (masculinized) charge and call to action becomes no less than the heroic rescue of the imperilled British Columbia.

The 'captivity narrative' (Schaffer 1991) at work here – of defenceless, feminized space whose boundaries require protection from 'the commingling of blood' (Canada 1922: 1518, 1522, 1524, 1529) – structured the discursive terrain of province and nation-building. It created and appropriated the bounded notions of 'British Columbia' and (writ large) of 'Canada', drawing on specific codes of race, femininity and masculinity within Canadian culture. These were the constitutive cultural and political resources whose interaction we need to glimpse in order to appreciate how the 'ideological work' (Poovey 1989) of nation-building actually got done. That those codes were powerful and persistent is doubtless also the case, as witness the 'race hygiene' debates of the 1930s (see Anderson 1991: Ch. 4) when there was very explicit concern in Canadian policy circles about the 'mongrelization' of White purity by Chinese 'penetration' (see also Stepan 1991). It follows that we need to resist any reductive impulses to distil a governing logic of binary (race or gender) relations to more fully understand the irreducibly *broad* social projects entailed in the construction of colonialism's cultures.

SEX, VICE AND CHINATOWN: RACE, GENDER AND PLACE

Vancouver's texted place called 'Chinatown' was crafted out of a repertoire of images whose racialized *and* gendered content enhanced their cultural appeal and political effect. In this section I would like to elaborate some of the gender-silences in my earlier race analysis with reference to the field of knowledge surrounding Chinatown in the period between approximately 1880 to 1930. During that time period, it is evident that the concept of 'Chinatown' harnessed racialized images to its service that were already deeply gendered (see Ware 1992) and which drew on more than a more narrowly conceived Orientalist field of knowledge. By revisiting a series of illustrations from the

Figure 12.1 'Miss British Columbia: "Will Sir Wilfred Laurier never touch that button?"' (*Saturday Sunset*, 4 September 1907)

Vancouver press earlier this century, I hope to expand the interpretive grid I cast on such historical materials and illuminate the discursive network within which operated race, gender, and sexuality languages and practices.

I previously argued that, of all the things that might have been said by early White Vancouver residents about Chinatown, only those aspects that fitted the racial categorization became filtered into the neighbourhood construct. In the early twentieth-century cartoons from Vancouver's *Saturday Sunset* (Figures 12.2, 12.3, 12.4), we see the nature of the material out of which Chinatown was ideologically constructed. Local knowledge drew upon the presumed proclivities for opium, gambling, sexual exploitation and over-crowding of that abstracted figure 'John Chinaman' to produce interlocking registrations of vice, mystery, danger and disease. For those seeking to render the place and its people eternally alien, a label into which could be assimilated all the things that Europeans sought to deny or repress in themselves, served a persistently useful function. Certainly the label became more than a package of (derogatory) meanings, words and texts. As I demonstrated in my earlier work, the Chinatown concept triggered and justified harassment campaigns for many decades as part of the state's management of ethnic pluralism.

One can go further, however, than arguing that European conceptions of 'a Chinese race' was the primary modality governing law enforcement practices in Chinatown. There seems also to have been constructed around Chinatown a deeply gendered 'moral panic' (Cohen 1972) that served to legitimize not just White Canadian intervention but a historically specific form of masculinity and moral guardianship. In the 1908 illustration 'Vancouver Must Keep this Team (Figure 12.2), civic guardians Chief of Police Chamberlain and Magistrate Alexander – icons of the law enforcement arms of the state – are clearing their heroic path through Vice-town. It is a site where 'difference' is being scripted as 'danger' to culturally dominant norms; a place that the 'axe man', Deputy Chief of Police McLennan, sought tirelessly to 'tame' (*Province* 16 July 1913). And that the norms these men felt moved to police were themselves crafted out of gendered (and heterosexist) material is plainly evident in the 1907 cartoon 'The Unanswerable Argument' (Figure 12.3). The ideal of Canadian, suburban, civilized, family life is here set up in opposition to, and at risk from, the pathologized modes of living on Carrall Street.

Although the competition presented to the White working class by cheap Chinese labour was a persistent theme within early Vancouver's colonial discourse, it was the scenario of sexual liaisons between Chinese men and White women which was seen as the most threatening violation of all. Indeed nothing served to congeal stereotypical knowledge about Chinatown more securely than the emblematic activity of John Chinaman's predation. So while in certain instances, Chinese could be cast as 'a feminine race', to use the words of Royal Commissioner Dr Justice Gray in 1885 – 'docile' and well suited to the menial labour of railway construction (Canada 1885b: 69) – they could, to serve other purposes, be masculinized and construed as energetic pursuers of White women (see also Back 1994). The contradictions within the languages that constructed Chinese as alien were complex and do

Figure 12.2 'Vancouver must keep this team'
(*Saturday Sunset*, 1 February 1908)

Figure 12.3 'The unanswerable argument'
(*Saturday Sunset*, 10 August 1907)

Figure 12.4 'The "foreign mission field" in Vancouver'
(*Saturday Sunset*, 10 October 1908)

themselves point to the irreducible diversity of classed, raced, gendered and sexualized resources upon which colonial discourses drew.

There were other double standards. Whereas sexual relations between Chinese women and White men were rarely discussed (except to fuel alarm about the transmission of syphilis), the possibility of sexual relations between Chinese men and White women was deeply troubling (if also, perhaps, titillating) and supplied much discursive material for Chinatown's image-making. The 'moral blight' that was Chinatown certainly set a pressing agenda for Christian missions in early Vancouver. In the 1908 cartoon 'The Foreign Mission Field in Vancouver' (Figure 12.4), Chinatown's opium dens are constructed as the natural habitat of the lascivious John. Inside them, White women – passive at the hands of the inscrutable Oriental – are induced to commit 'amoralities', in the words of many a civic official. As occurred in other 'inter-racial' settings, the scenario is made to stand as the most profound of violations, and it worked in a few ways: first by contacting the generalized fear, beginning at the point of immigration, that 'Canada' faced a threat from close contact with outsiders. Proximity of 'races' within the private sphere could thus also be construed as 'perilous' by image-makers such as Attorney-General Mason and the press, both of which played wickedly on the notoriety of 'Vice-town' in 1924 when a Chinese domestic ('China boy') allegedly murdered his employer, Janet Smith, of the high-income district of Shaughnessy (see *Province* 13, 28 Nov., 5 Dec. 1924; also Lee 1990: 65–9). This was one of a few occasions when alleged murders of White, wealthy women by Chinese servants was used to transform difference into 'danger' and to justify the enforcement of boundaries between Chinese men and White women (see also the controversy surrounding the death of Mrs Millard of the West End in 1914 in Anderson 1991: 116).

Second, the fantasies and anxieties surrounding sexual relations between Chinese men and White women fed into other cultural discourses. Within such discursive 'fellowships' (Foucault 1972) circulated languages of: presumptive heterosexuality (according to which White women were the exclusive preserve of White men); racialized manhood (such that the 'bestial negro' and 'wily Oriental' could be rendered lustful primitives); White femininity as an innocent and vulnerable essence; and of women as Othered objects, servicers of male bodily needs and desires. Like their insatiable pursuers, White women were also closer to nature than the rational, controlled, White male. Small wonder, then, that sex between Chinese men and White women became inscribed as the ultimate moral and political transgression. Not only did it compromise racial boundaries, it threatened White, male property. Safeguarding the virtue of White women thus became dignified as a prerogative, and was very often the pretext that law enforcers used in targeting Chinatown. In so doing, White men exercised not just their sense of race and gender supremacy, but also their power of definition over the criteria of normalcy in sexual conduct.

The sexual politics at work in civic missions to Chinatown trouble the binary frame of race analyses that assume an essential opposition of interests between 'Whites' and 'Chinese'. For one thing, the representations and practices surrounding Chinatown originated in social relations that included gender and sexual orientation, suggesting there was nothing unitary in the position of racial privilege (or, as I shall later argue, of racial subordination). White women's inferior positioning relative to White men, together with the privileging of heterosexist masculinity in Canadian culture, informed the very moralities that grew up around the 'race' question. Indeed they compromise any narrative characterizations of social relations that might collapse gender, sexuality and class arrangements into a larger governing conception of 'race' domination. Nor was it a case of a simple layering of race with gender meanings (as might be implied by linear and mechanistic 'additive models' of oppression; see Sacks 1989; Spelman 1988). The point underlined by the substantive discussion is that racist knowledges had gender and moral codings relating to family, sexuality, marriage and residence embedded within them, just as discourses surrounding gender, sex, citizenship and family life relied on race meanings for their cultural integrity. It follows that race identities cannot be decontextualized and separated off analytically or politically from the constitution of other identities and axes of power. Each division is practised in the rhetorical and interactive context of others. The representational practices surrounding Chinatown thus bring into view the insinuation through each other of the multiple hierarchies that underwrote early Vancouver society.

CHINATOWN REORIENTED: ALTERNATIVE SUBJECT POSITIONINGS AND STORIES

The contribution of racialized and gendered discourses to ethnic relations in early Vancouver might seem to have been so decisive as to support readings of colonialism as a pervasively efficacious venture. If the making of a *British* British Columbia was a relatively influential project, however, it was not a unitary one evolving from a singular source of ethnic superiority. We have seen that its sources were multiple and differentiated. It is also the case that, for all of colonialism's power at certain times and in particular places, the management of ethnic difference was no neat process of imposition. Rather it entailed struggle, and was often fragmented and frustrated by debate, contradiction and resistance by those it subordinated.

Analysis of colonialism's operation can also be constructed from a range of vantage points. Indeed the perspective of the elite White master-subject is possibly only legitimized by accounts, such as appear in the first part of this chapter, of the simultaneity of race, gender and heterosexist oppressions in colonialism's extension to British Columbia. Thus while it is important to continue to illuminate the differentiated sources and forms of power under colonialism, it is also helpful to puncture the binding grip of *a* (master-) story of the inter-ethnic encounter – a grim tale, that is, of inexhaustibly coherent control on the part of a privileged Anglo group. In what follows, I seek in a most preliminary way, to diversify the Chinatown problematic by opening up the story-field to alternative subject and speaking positionings.

Although there can be little doubt that White Vancouver women were often complicit in the practices that marginalized Chinese in that city's early history, there were moments when White women broke ranks with White men and formed alliances that undercut the stable fixings of racialized boundaries. A more nomadic style of story-telling to the linear mode of *Vancouver's Chinatown* (1991) – one that weaves narrative threads through scattered moments – illuminates such apparent ruptures in the Orientalist logic of Occident versus Orient. To that end, vignettes that I earlier framed in race terms (see Anderson 1990; 1991: 116–20, 158–64), can be recast to disrupt logics of race complementarity and to highlight the possibility of political alliances that cut across racialized identities.

In 1920, the Vancouver city council decided to impose a hefty $100 licensing fee on vegetable peddling, a trade almost wholly dominated by Chinese in early Vancouver. The move on the part of council was undertaken out of support for the powerful Retail Merchant's Association which had grown concerned about the inroads being made into its business by the itinerant vendors. The pedlars – disinclined to accept the fee – decided to enlist the support of their clients as well as the Chinese ambassador. Interestingly, on this occasion, some 5,000 Vancouver women were more interested in avoiding long shopping trips to the city market than endorsing the vendetta of the Retail Merchant's Association. The women signed the petition in support of their Chinese produce suppliers and against the White male retailers who, in race readings of social relations, are cast as their compulsory partners.

The incident wasn't the only occasion when there were alliances *across* racialized boundaries that are written out by a logic of binary opposition. There were other moments of vulnerability in dominant discourses surrounding Chinatown. Fifteen years later, in 1935, Chief Constable W. Foster found cause to implement British Columbia's Act for the Protection of Women and Girls, specifically in Vancouver's Chinatown. That Act had been implemented back in 1919 out of fear for women's 'moral safety' in Chinese restaurants throughout British Columbia. In Vancouver, Foster argued that contact between Chinese men and White women was being set up inside restaurant booths and that after working hours, women would go to Chinese quarters where 'they were induced to prostitute themselves and immorality would take place' (cited in Anderson 1991; Ch. 5). This was no trivial matter for the retired colonel, and between 1935 and 1937, Foster and his 'moral reform squad' comprising Mayor M. Miller, License Inspector H. Urquart and City Prosecutor O. Orr set about banning White waitresses from Chinatown cafes by cancelling the licenses of businesses employing them.

The vendetta against Chinatown cafes met with angry resistance from Chinese owners of the restaurants, including the powerful and wealthy president of the Chinese Benevolent Association, Charlie Ting. Perhaps the most revealing challenge, however, came from the women themselves, who in 1937 marched to City Hall to protest their dismissal by, in the words of one woman, 'the self-appointed directors of the morals of women in Chinatown'. The women were quite prepared to articulate the specificity of their experience as workers and defend their right to choose their employers and place of work. Certainly the women defied the image of the passive object of desire that we have seen was so useful in dignifying earlier male missions in Chinatown. One waitress, Kay Martin, told the press she would 'much prefer working for a Chinese employer than for other nationalities'. Another stated that 'if a girl is inclined to go wrong she can do it just as readily uptown as she can down here'. Another noted 'our bosses are honourable men who know that we must live'.

If it was the adversity of living conditions during the Depression that brought about the womens' defence of their employers, the action nonetheless upsets readings of relentless race polarization. Such interpretations risk flattening the experiences of White women as subjects, deducing them from the (putatively) immobile and deterministic position of race power. Yet as we have seen, 'Whites' were not always and necessarily fated to dominate. The apparent coherence of racism gives way before such evidence, which, while necessarily brief here, highlights the mutable configurations, crosscutting constituencies and contingent authorities out of which social relations are made.

Similarly unsettling of dualistic race readings are the distinctions of class, gender, ethnicity, generation, language and so on that pluralized Chinatown as deeply as they did White Vancouver. If the likes of Charlie Ting became, for a time, an ally of White women in Vancouver, those of his class may have been less 'honourable' in the eyes of the Chinese rank and file workers of the enclave economy in Vancouver's Chinatown. Many scholars have identified a socioeconomic pyramid in the district, at the apex of which stood a tiny minority of men of capital who were some of the wealthiest individuals in early Vancouver (Wickberg et al. 1982; Yee 1988). The liability implied by the racial category 'Chinese' may well have been the asset of certain merchants who in Chinatown had a vulnerable and captive labour force at their disposal. This bloc of workers, unprotected by White unions, often laboured under punitive contracts for their Chinese bosses. There were also many unpaid workers, including women, who worked long hours sewing buttonholes and doing much of the handwork for Chinese tailors (see Adilman 1984). The experience of those workers was shaped by their subordinate status in an array of dualities, not least of gender. The women prefigured today's sweat-shop workers whose notoriously exploited labour in other North American Chinatowns, such as New York City, tells of ongoing class and gender antagonisms which have only recently prompted agitation for reform on the part of Chinese women workers (Kwong, forthcoming).

Such gender-differentiated relations within the racialized category are erased by characterizations of the universally subjugated 'Chinese'. Not only do they suggest different racisms for different groups of 'Chinese' (see Satzewich 1989), they also lead us to consider the possibility that other oppressions – quite apart from the relation that places Whites in a determin-istically antagonistic relationship to Chinese – might have been as decisive in shaping their everyday experience. Not all the realities and aspirations of the lives of Chinatown's residents would have been exhausted by the fact of racial subordination, as the growing number of fictionalized accounts by Chinese Canadian writers of life in Vancouver are beginning to reveal (Chong 1994; Lee 1990).

CONCLUSION

This chapter has sought to confront the tensions raised for analysis by the intersection of axes of socio-spatial inequality. I have sought to undo the privileging of *racialized* positionings – European versus Chinese – by foregrounding the gendered meanings and practices that at times reinforced, and at other times, disrupted those categorizations and relations. In the post-structuralist spirit of a more 'distrustful analysis' (Bottomley 1991: 108) that eschews the search for unitary subjects and singular explanatory frameworks, I have attempted to decentre the authorial paradigm of 'European hegemony' charted in my own *Vancouver's Chinatown* (1991). Without discrediting the case for racism's force, malleability and resilience in Canadian culture, or indulging a naive postmodern embrace of endlessly infinite identities, I have here tried to demonstrate that the cultural field is created and fractured by a range of social relations and subjectivities whose mappings invite what Pratt (1993) has called a 'restless story-telling'. By emphasizing the different centres of cultural authority surrounding race, gender and sexuality, and the invariant political alliances surrounding those idioms of identity and power, the chapter has sought to disrupt modernist notions of undifferentiated subjects, root causes and fixed trajectories. It follows that the Chinatown story-field might effectively be further opened up to re-tellings from the vantage points of the district's residents, themselves multiply and fluidly positioned in relation to each other and the wider society.

DISPLACING THE FIELD IN FIELDWORK

Masculinity, metaphor and space

Matthew Sparke

[A] point that needs to be emphasized here is that certain spatial metaphors are equally geographical and strategic, which is only natural given that geography grew up in the shadow of the military. . . . *Field* evokes the battlefield.

> (The editors of the journal *Hérodote*, in Foucault 1980: 69)

[O]ur relationship to the 'field' itself must be problematized; that is, we must recognize that the field is constructed through power relations that define academics and the people and places we study.

> (Staeheli and Lawson 1994: 97)

'[T]he principal training of the geographer', once declared Carl Sauer (1956: 296), 'should come, wherever possible, by doing field work.' Numerous other famous men in the discipline have said the same. Moreover, as Gillian Rose has recently argued, they have said so simultaneously lauding it as a tough and heroic activity, as 'a particular kind of masculine endeavor' (1993: 70). All the while work in the field has been sanctified as a character-building rite of passage into a world described as real, the field itself has been feminized, cast as a seductive but wild place that must be observed, penetrated and mastered by the geographer who, having battled with it, revelled in it, and, in the end, triumphantly risen above it, returns to the academy his education complete, his stature assured and his geographical self proven, definitively, *his*. Such a sentence, though, however lengthy, summarizes the masculinity of fieldwork rather too roughly. To be sure, its exclusively masculine pronouns, heroes and assumptions are commonly announced with authority by the father figures of geography. However, there are also multiple, concrete examples of how its seminal logic has become *dis*seminated, which is to say, performed and thereby transformed by other, less manly, geographers.[1] As Jennifer Hyndman (1995) has argued, 'many feminist and other geographers do fieldwork precisely to critique, deconstruct and reconstruct a more responsible, if partial, account of what is happening in the world'. There are, then, possibilities for renegotiating fieldwork for the better, and my aim in this chapter is to trace the complex web of practices and metaphors which, while consolidating the field as a site of masculinist work, have begun nevertheless to be rearticulated as a politicized location for endeavours with more emancipatory promise.

An apologist for traditional fieldwork might well dismiss my initial outline of a critique by arguing that the metaphorization of the field as feminine proves nothing. In the first section of the chapter I seek to address such criticisms by connecting my discussion with some of the more general debates currently circulating about spatial metaphors. In particular, I draw on writing by Cindi Katz (1993) and Neil Smith, who argue that spatial metaphors become politically problematic when they introduce a fixed and fixing notion of *absolute space*.[2] Masculinist geographical imaginations of the field as a feminized, separate and containable space, work with, or so I will argue, just such a dehistoricized conception of absolute space. It is also a conception which, to recall the argument of the French geographers interviewing Foucault, owes a great deal to the strategic geo-politics of the military. It evokes the battlefield; and, just as military conceptualizations of space are connected directly with the violence of war, so too do masculinist formulations of the field of fieldwork have quite literal implications.

Yet it would be ridiculous to assert that every plan for a field trip somehow amounts to a virulent call to arms. Care is required in distinguishing between the overarching dynamic of disciplinary violence and the heterogeneous ways through which different fieldworkers renegotiate its influence. For this reason, the later sections of this chapter move beyond the question of militarist *cum* masculinist metaphors in order to take up the substantive task of problematization outlined by Staeheli and Lawson (1994). Their suggestions, along with the other feminist discussions of fieldwork published in a special methods section – 'Women in the Field' – of *The Professional Geographer* (46 (1): 1994) move the debate on to how to actually think and do research that avoids inflicting violence on those who are researched. In doing so, they not only problematize the masculinism of fieldwork, but simultaneously displace the meaning of the field itself.

Ultimately, the displacement of masculinism raises a question about the geographers who generally find it easiest to take masculinity for granted: men, and, in particular, straight men. For those like myself who would like to contribute to, rather than take over, attack or otherwise oppose feminist work, it means at the very least more self-reflexivity. More generally, I feel that the lessons of feminist critique demand of men a willingness to read and discuss a whole range of critical work in a spirit of responsibility. Learning from feminist work in this way means that self-reflexivity need not, indeed, must not be limited to bold identitarian announcements of the 'as a straight white man' variety. Such 'as-a-ism' seems only to provide a mantra of relief from the more detailed and difficult task – the 'perverse' task, as Sandra Harding (1991) describes it – of examining the complex *contradictions* constituting one's positionality. Thus, in the final part of the chapter I bring the questions introduced by feminist critique to the specific contradictions involved in the conflicting ways masculinity and shared working experiences shaped my own fieldwork interviews of temporary workers in Vancouver, BC.[3] This, then, is not an attempt to pretend that I can speak from a woman's position – even if it could be so singularly defined. Instead, my argument represents the situated knowledge of a man interrogating the masculinity of fieldwork by turning to feminist work.

FROM THE FARAWAY AND FEY TO THE FIELD AS A SPACE OF BETWEENNESS

Concluding her introduction to her book *Technologies of Gender*, Teresa de Lauretis describes an arc of feminist critique that traces what she calls 'a movement from the space represented by/in a representation, by/in a discourse, by/in a sex gender system, to the space not represented yet implied (unseen) in them' (1987: 26). De Lauretis's more specific focus is on the visual arts, but, in this first section of the chapter, I will argue that the particular space of the field can also be traced through the same arc of representation and implication. To begin with, I outline the dominant masculinist demarcation of a feminized field. As a spatial concept-metaphor that has been shaped by/in worldly practices, I argue that the field has become normalized as a disciplinary concept, and, as such, has had world-constitutive effects. However, the way in which these effects have depended on the particular practice of fieldworkers tracking between field and academy has done more than serve as the adventurous stuff of manly myth-making. I shall suggest that it has also been its undoing. As new and increasingly critical scholars have come to retrace the back and forth route, they have brought into crisis the very distinctions – between researcher and researched, the near and far, aesthetics and politics, scholarship and practice, mind and body, masculine and feminine and so on – upon which it historically rested. In doing so, they have *rearticulated* – in the double sense of both 'rejoining' and 'repeating out loud' – a space of between-ness that was, I will argue, precisely the space that was previously assumed but disavowed, 'a space not represented yet implied (unseen)' within the older narratives of manly exploration.

The fielding of the field

In various Western discourses 'field' is associated with agriculture, property, combat, and a 'feminine' place for ploughing, penetration, exploration, and improvement. The notion that one's empirical, practical activity unfolds in such a space has been shared by naturalists, geologists, archeologists, ethnographers, missionaries and military officers.

(Clifford 1990: 65)

Clifford's summary provides a useful starting point for a genealogy of the field. Geographers too could clearly be added to his list of researchers sharing the space, and the questions he asks of anthropology are for the same reason just as relevant to geography: 'What commonalities and differences link the professional knowledges produced through these "spatial practices"? What is excluded by the term "field"?' (1990: 65).

I will return to this metaphorical yet material notion of spatial practice shortly, but one of the commonalities here is undoubtedly that of masculinity. The agent of action, whether it be of War, of God or of Science is assumed to be Man. Woman can symbolically serve as his helpmate, but it is he who proves and improves himself through a mastery of the field. By way of a corollary, a possibility that is systematically excluded by these gendered networks of power and knowledge is that of women acting as agents of knowledge. Instead, they are repeatedly seen – and the visual verb is not coincidental[4] – as just part of the field. Fieldwork is in this way symptomatic of the more general disciplinary tendency in geography described by Rose (1993: 9), a tendency wherein femininity becomes the Other, the space above which and against which geographers of reason define their science, their art and, in these differing ways, their selves.

Considering the actual history of geography it is important to remember that associations like those outlined by Clifford constitute more than just verbal or writerly coincidences. Anna Skeels's detailed historical examination of Sauer 'in the field' has indicated that his own conceptualization of field education was as a certain *rite de passage* into manhood, a 'formation of the field*man* from the "boys"' (Skeels 1993: 89). Reworking such views, Richard Symanski's (1974) notorious fieldwork into brothels in Nevada might well have been an extreme and idiosyncratic example, but, as Barbara Rubin (1975) pointed out in her critique at the time, it illustrated quite clearly the ways in which a masculinist scholarly attitude had decisive consequences for the sexualizing of actual research and the objectification of actual people. Likewise, when David Stoddart (1986: esp. 143–57) – to retain one of Rose's main examples – champions fieldwork as the hardy stuff at the heart of the discipline, when he commends Sauer for his insistence on the geographical gaze (1986: 147), and when he connects all of this to his own attempts at 'making sense of nature' (1986: ix), he illustrates a continuum between thoughts, words and deeds. It should be remembered, though, that these forms of flamboyantly unabashed claim take place within a much wider network of social practices such as those alluded to in Foucault's interview with the geographers. Take as a particular moment of condensation Stoddart's plea to the discipline to have pride in the history of fieldwork as 'a record of achievement at the farthest ends of the earth'. 'Let us', he opines,

> salute with Conrad 'men great in their endeavour and in hard-won successes of militant geography; men who went forth each according to his lights and with varied motives . . . but each bearing in his heart a spark of the sacred fire'.
>
> (Stoddart 1986: 157)

Here the associations noted by Clifford are articulated with grandiose aplomb. Masculinity, militarism, imperialism and science all come explicitly together in a fantasy of fieldwork in faraway lands. While, as Derek Gregory argues, their 'modalities of power lie beyond the compass of [Stoddart's own] account', 'their main thrust' remains stridently, albeit nostalgically, clear (1994: 32, 20). More than just the exploits of lone men, then, it is within the burning context of what Conrad fetishized as 'sacred fire' – namely, within the violent masculinist discourses linking science, empire and exploration – that traditional fieldwork has had its foundation and force. Stoddart's heroic rhetoric simply makes the connections more apparent.

The fact that the masculinity of fieldwork has been fashioned in the context of general, indeed, imperial systems of knowledge production should warn against any personalist attempt to 'blame' the masculinist construction of the field on especially flagrant individuals. Given that the problem is more general than individual intentions, so too must be the critique. As Rose (1993: esp. Ch. 4) indicates, the macho model of men entering the field is also over-determined by the feminization of nature in science, and the privileging of the knowledge of the masculine gaze. Insofar as geographers have inherited a place in these traditions, they need to reconsider the masculinist and imperialist arrogance of the discipline at large. It seems inadequate for a commentator like Denis Cosgrove to simply poke fun at what he calls the 'hairy-chested feats of scholarly endurance' (1993: 516) advocated by those seeking disciplinary redemption in fieldwork. In this case, the writers he is responding to had argued for a form of field stamina involving 'not merely physical exertion but also the intellectual discipline that comes from engaging in ethnographic research . . . in non-English-speaking settings' (Price and Lewis 1993: 9). Against this type of claim, a more adequate critique would have had to address the overdetermined masculinist and Anglo-centric production of neo-colonial 'intellectual discipline' itself. Perhaps eulogies to fieldwork at 'the farthest ends of the earth' do sometimes express a 'muscular disdain for the fey and metropolitan' (Cosgrove 1993: 516). But coming from Cosgrove, a man who elsewhere affirms an 'epicurean relish for what Stoddart calls "lands of delight"' (Cosgrove and Daniels 1989: 179), the criticisms appear rather superficial. The point surely is that whether it is done with muddy boots or intellectual discipline or both, whether as a 'recording science' or 'performing art' (1989: 171) or both, fieldwork remains imbricated within masculinist modalities of power. These may, as Rose (1993: esp. Chs 2, 3) suggests, be understood as different genres of masculinity, some scientistic and others aesthetic,[5] but the construction of the field as either faraway or fey remains nonetheless the construction of an Other: the field as something to be looked at from above, to be struggled with and enjoyed on the ground, and always, in the end, mastered.

The systematicity of it all means that breaking away from a masculinist approach to fieldwork is often easier said than done. Cosgrove might be argued to come closer when he and Daniels (1989: 179) note how relevant 'the metaphor of the mirror' is for describing the performance of fieldwork. How-ever, the notion of the field functioning as masculine geography's self-consolidating, speculative Other is still left unexamined by the two men. Instead, the metaphor of the mirror becomes for them 'congenial' not critical, and thus ultimately appears, in David Matless's words, 'curiously conservative' (1989: 182). In anthropology too, the difficulty which men seem to have of problematizing the masculinity of fieldwork is also clear. Paul Rabinow, for example, in a book that otherwise problematizes so much, fails to deal at all critically with the masculinity of his fieldwork in Morocco (Rabinow 1977). Instead, in a section of the book that relates a growing intimacy with the place, and a growing rapport with a male informant and his 'roguish circle' (1977: 61), he writes what might well be called a narrative of penetration, culminating in his having sex with a so-called 'Berber girl' (1977: 69). If it were possible – which I do not think it is – to disconnect what happened in this instance from what continues to happen, no doubt sometimes violently, to young women around the world in the context of white, male fieldwork, the section could at least be read like the rest of the book as a usefully frank discussion of the feelings of the anthropologist. But even as such, the feelings are not problematized as masculine feelings, and there is no effort made by Rabinow to examine the event as a practical embodiment of the more general power relations privileging a white American man entering, studying and, in his own words, having 'sensual interaction' (1977: 65) with the objectified Moroccan landscape.

A factor that emerges as a defining characteristic of masculinist framings of the field is the capacity assumed unquestionably by the fieldworker to be able to leave. Very few of the people who are researched are ever able to even pretend to do the same. Their reflections on fieldwork are rarely written down, let alone circulated through the academy. Meanwhile, as the people and places visited by the likes of Rabinow, Stoddart and Symanski are fast transformed into mute objects before an assimilative academic gaze, the fieldworkers themselves become ensconced in the academy with all the authority due to scholars who have braved the field and returned. In this sense, the double movement of going *There* only to come back *Here* constitutes the elemental spatial practice, the primal *fort/da*,[6] at the very core of masculinist fieldwork authority.[7] In the next section, I examine how this spatial practice is itself secured through a masculinist metaphorization of the space of the field.

The field of masculinist in-fluence

The trope of space must be consciously analyzed in order to evaluate its in-fluence.

(Kirby 1993: 188)

In order to understand how the spatial metaphor of the field is caught up in the construction of a fieldworker's authority it is necessary to consider what Katz and Smith (1993: 68) refer to as 'the interconnectedness of material and metaphorical space.' This interconnectedness is important to remember not only because it opens the question of how spatial metaphors such as the field come to shape social life, but also because it serves as a caution against any interested attempt to fix where metaphoricity ends and conceptualizations of materiality begin.[8] As Dominick LaCapra's (1980) deconstruction of Ricoeur's work shows, there is an implicit violence done in any such theoretical project to legislate the meaning and scope of metaphor once and for all.[9] As an alternative it is possible to take Derrida's own approach and consider how the interrelating flux of metaphors and concepts is abbreviated through philosophically or, as in the present case, politically interested moments of closure (Derrida 1982). Elsewhere I have described how this process of enclosure can effectively dehistoricize, homogenize and contain the reference of spatial names and metaphors so as to produce what I called anemic geographies (Sparke 1994b). Here, the case of the masculinist field presents another anemic geography. However, in contrast, for example, to the racist metaphorization of Africa as 'Dark Continent' critiqued by Lucy Jarosz (1992), it is one which takes place at a more personal and yet generalized scale. It privileges the individual fieldworker by securing the field wherever and whatever it is as separate and contained. It is this specific moment of closure that needs to be examined in terms of the irretrievably entwined relations of 'material and metaphorical' space.

Katz and Smith (1993: 75) argue that '[s]patial metaphors are problematic in so far as they presume that space is not'. They go on to outline how this form of presumption becomes possible when space is conceived along the geometric lines of what they call *absolute space* (see note 2). It is a conception that effectively *makes* space seem unproblematic by reifying it, removing it from the dynamics of its historical production, and naturalizing it as a timeless and measurable given. I think it is precisely this same absolutist way of imagining space upon which the fixing of the field is predicated. Pinned to the depthless horizon of absolute space the field can easily be presented as an unproblematic domain lying outside the academy, a space, then, that dissembles its own complex and all too academic production. As such, the presentation of the field draws on a tradition in which space is seen after Euclid and Descartes as 'geometrically divisible into discrete bits' (Katz and Smith 1993: 75). To borrow a definition from Martin Jay, 'as [a] spatial metaphor . . . [the] field [thus] tacitly assumes a synchronic entity to be surveyed or mapped as a structural or relational gestalt' (Jay 1990: 312).

For Katz and Smith the connection between the field and the problematic of absolutization is clear. The one serves as a metaphor for the other. Absolute space, they say, 'refers to a conception of space as a field, container, a co-ordinate system of discrete and mutually exclusive locations' (1993: 75). However, this raises the question of why the field should provide such an amenable example of absolute space. How has its fixity and finitude endured through all the turbulence of fieldwork over the years? Two closely connected reasons are, I think, quite clear. The first concerns the sexuality of vision, and the second, the hegemonic ways in which modern Western thought has made the world seem picturable in the first place.

The field has been able to share and lock into the compartmentalization implicit in the *logos* of absolutized space by serving simultaneously as the feminized object of the masculine gaze, and the pictured place of communion with the actual and factual (see Rose 1993: esp. Ch. 4). This sort of place of communion has itself been established, or as Timothy Mitchell (1988) has put it, 'enframed' through an episteme of Cartesian dualisms. Like the *mental* versus *material* 'everyday metaphors of power' Mitchell seeks to displace, the no less powerful *academy* versus *field* dualism is thus supported by the hegemony in Western thinking of distinctions between meaning and reality, structure and practice, mind and body (Mitchell 1990). Coming together with the masculinism of the academy and the feminization of the field, these dualisms enable the ongoing compartmentalization of the field as a disciplinary version of what Elizabeth Grosz (1990) has called a 'body-map'.[10] Such a corporeal cartography would seem in this case to be coordinated so as to demarcate, contain and thereby *in*corporate all that the masculine disavows. As a result of such corseting, the field's meaning becomes encased, even, one might say, incarcerated. Seemingly barred from metaphorical movement, it is consolidated more as a concept, a durable, everyday and taken-for-granted embodiment of absolute space. And it is through this quotidian hegemony that it has functioned again and again to guarantee a place to which researchers can go secure in the knowledge that they will always be able to leave.

The comings and goings from a feminized field of masculinist work can now be better understood as a spatial practice coordinated through a spatial metaphor turned absolute, routinized concept. As such they have quite literal implications: 'dead literal' to use Donna Haraway's more poignant words (1989: 58). The examples of Symanski's studies, Stoddart's rhetoric and Rabinow's reflections provide only a limited indication of this, but Haraway's own incredibly detailed critique of primatology shows just how violent, even lethal, masculinist constructions of the field can be. An essay, for example, which follows Carl Akeley into the field documents exactly the destructive intersection of manly adventure with science and art in self-styled heroic work that aimed with both gun and camera to master the field of African nature (Haraway 1989: 26–58).[11] However, Haraway does not leave the story there, and while much of the rest of the book examines the racism, anthropocentrism and familial-sadism structuring primate studies, her final section opens the possibility of feminist renegotiation with a subtitle that speaks of primatology as a 'Genre of feminist theory' (1989: 278–383; see also Haraway 1991b). It is this same possibility for renegotiation that I seek to highlight next as I turn to the work of feminist geographers. As Heidi Nast suggests in her introduction to the 'Women in the Field' papers, this work can evoke a very different field, one that 'is not naturalized in terms of "a place" or "a people"; [but] rather . . . located and defined in terms of specific political objectives' (Nast 1994: 57). (Re)placing the field of fieldwork within such a politics of location does indeed seem to promise what Haraway (1989: 288) calls 'the possibility of new stories not strangled by the same logics of appropriation and domination'. As feminist research that is still empirically grounded, it also highlights how 'the intervention must work from within, constrained and enabled by the fields of power and knowledge that make discourse eminently material' (1989: 288).

Rearticulating the blasphemy of between-ness in fieldwork

> Under contemporary conditions of globalization and post-positivist thought in the social sciences, we are always already in the field – multiply positioned actors, aware of the partiality of all our stories and the artifice of the boundaries drawn in order to tell them.
>
> (Katz 1994: 66)

To paraphrase de Lauretis (1987), my argument thus far has suggested the following: that the traditional space of the field has been presented by/in specific disciplinary systems of representation – call them scientific data retrieval and interpretative observation; that these have been upheld by/in an organization of power and knowledge comprising a back and forth spatial practice of masterful study – call it 'going there', 'being there' and 'leaving'; and that this form of spatial practice has been fashioned by/in a system of sex-gender – call it heroic masculinism. Following de Lauretis, the critical move that needs now to be made is one that asks what spaces are assumed but yet disavowed through this formulation of the field. If the absolutization of space enables the field to be assumed as securely distant and contained, what 'active and stirring' spatial relations are concealed beneath such a projective palimpsest? What is the spatiality 'not represented yet implied (unseen)' by the masculinist fielding of the field (Derrida 1982: 213)?[12]

Critics such as Katz have already engaged in exactly this form of questioning. Insisting, in her words, on how 'we are situated and bear responsibility for interrogating our positionings' (1992: 504), they have moved away from the Olympian high ground of objectification and Cartesian distinction. Instead, by drawing attention to subjects that previously went neglected, they have highlighted the geographies of power that, in the words of Staeheli and Lawson, 'define academics and the people and places we study' (1994: 97). This is a radical problematization of the field. Rearticulating what the absolutization of space previously kept apart – and thereby intangible, inaudible and unexamined – feminist critique has displaced the dualism of field and academy, replacing it with the more grounded yet dynamic notion of fieldworking in what Kim England calls 'the world *between* ourselves and the researched' (1994: 86). This is not, she notes, the same as conducting fieldwork 'on the unmediated world of the researched' (1994: 86). Nor is it another reincarceration of the field as radically Other. Instead, it is a repetition of fieldwork with a difference, a form of blasphemous empiricism that admits to subject positions and, hence, research positions constituted materially and interpersonally in what Katz rearticulates as 'spaces of betweeness' (1994: 72).

By describing the attention to interpersonal positioning in feminist fieldwork as blasphemous I mean to invoke the same ironic mix of fidelity and dissent made manifest by Haraway. Such blasphemy, she says, 'has always seemed to require taking things seriously', but it also 'protects one from the moral majority within, while still insisting on the need for community. Blasphemy is not apostasy' (1991c: 149). Feminist empiricism seems blasphemous in much the same way.[13] Frequently dissenting from high theory – both masculinist and feminist – it has had what Linda McDowell (1993) describes as a long and subversive history within geography. Faithful to the responsibility of representation through fieldwork, and insistent on the need for rigorous research that makes a difference for communities, it has also nevertheless blasphemed against the rules and rites of what to study, how, when and where. Pam Moss (1993), for example, describes how it has stressed qualitative over quantitative approaches, and, as Isabel Dyck (1993) underlines, this same stress on the knowledge produced through intersubjective communication has brought with it a dissenting reflexivity towards the gendering of research itself. It is this blasphemous empiricism that also provides a vital backcloth against which to discuss a point of seeming contradiction addressed in a footnote by Rose.

In the footnote in question Rose writes: 'I am not suggesting that women cannot undertake fieldwork only that its dominant style is a tough masculinity' (1993: 181). One way, perhaps, of thinking about the blasphemy of feminist fieldwork, is that it constitutes a critical renegotiation of this same tough masculinist style. Rearticulating what the style and accompanying spatial practice kept apart, the blasphemers raise complex questions about the relations of power linking the traditionally secluded spaces of academic life and fieldwork research. They respect what Clare Madge describes as the 'need to consider the role of the researcher in the research process' and, in doing so, bring about a 'boundary dispute' (1993: 296), a radical interrogation of the limits of the field. It is therefore a renegotiation that, far from abandoning the field, works instead by remembering geographers' ongoing positioning in between multiple overlapping fields. Katz's argument, 'I am always, everywhere, in "the field"' condenses this testimony to what before was only ever 'implied (unseen)' (1994: 72), and, for the same reason, presents probably the greatest possible blasphemy against the stylized, *fort/da*, spatial rituals of heroic fieldwork.

That they constitute blasphemy might also account for why reflexive statements about fieldwork and the academy have so rarely been uttered by male geographers. Certainly, when they have, the guardians of the sacred fire have descended. Allan Pred, for example, made what could well be read as a departure from the dominant penetrative model when he noted that:

> the distinction made between 'fieldwork' and other more everyday observations and experiences is but one manifestation of a general unwillingness to accept the fact that our 'professional' and 'non-professional' lives are not in dichotomous opposition to one another, but dialectically interrelated.
>
> (Pred 1984: 91–2)[14]

This was clearly blasphemy for Stoddart: writing by which, he reports, he was not much enlightened; an argument deserving only of derision, not profound, not physical, and, presumably, not tough enough (1986: 147).[15] Against this backcloth, then, the blasphemous achievement of the feminist methods papers in *The Professional Geographer* becomes clearer.

It may at the outset have seemed strange for me to turn to the example of the 'Women in the Field' essays as an example of feminist renegotiation with masculinist fieldwork. The critical attention of the authors is not turned directly towards the question of masculinity and, empirically, men and their immediate affairs are – in a rare move for professional geography – marginalized. Instead, England (1994) attends to the dangers of doing research as a straight woman about lesbian life in Toronto; Melissa Gilbert (1994) analyses her position of privilege in relation to the low-waged women she interviewed in Worcester, Massachusetts; Katz (1994) critically connects her work with children in rural Sudan and East Harlem, so as to ground her argument about the continuities of the field in the context of globalization; and Audrey Kobayashi (1994) discusses the 'coloring of the field', bringing attention to the dangers of ethnographic authority rooted in ethnicist or otherwise essentialist absolutism.

As Nast suggests in her introduction, the writers thus go beyond the literalist and, as such, exclusive textualism that has limited recent attempts at ethnographic reflexivity by a number of men in anthropology.[16] In contrast to some of these privileged performances of polyphony, the papers more immediately concern the politics of women doing fieldwork, and the problems facing feminist solidarity in the context of a racist, heteropatriarchal and capitalist society. However, it is in this very practical attention to the dangers of epistemic violence and appropriation that the writers make what I think is their most blasphemous break with masculinist fieldwork. Faithful to the project of research, they nevertheless put its gaze, its appropriative arrogance and its limits under critical vigil. Rather than claiming Olympian vision, they situate themselves as women within the relations of dominance through which they are privileged as fieldworkers. And, in doing so, they turn a contradictory position of being fieldworkers and feminists in geography – a position, to use McDowell's (1992) phrase, both 'inside and outside "the project"'[17] – into a politicized location for critique. Overall, then, as Staeheli and Lawson put it, 'these authors question the boundaries of "the field"' (1994: 97), and as they do so the whole reified map demarcating and separating field from academy comes to life, the spaces divided by its boundary-drawing becoming rearticulated in the blasphemous between-ness of interpersonal debate.

I do not want to romanticize or homogenize the interventions represented by the 'Women in the Field' papers. They introduce a range of different approaches, and while, for example, Kobayashi, insists on how there is 'more to gain from building commonality than from essentializing difference' (1994: 76), Gilbert, by contrast, provides a host of sobering reminders of just how difficult seeking such commonality can be. Nevertheless, there is a shared blasphemous impulse in the papers, and it is this – most especially, their attention to the partiality of situated knowledge, and their rearticulation of field and academy – that I would like to bring to a re-examination of my own fieldwork in Vancouver.

OF FIELDWORK, TEMPING AND THE POSSIBILITIES OF POSITIONALITY

We do not seek partiality for its own sake, but for the sake of the connections and the unexpected openings situated knowledges make possible.

(Haraway 1991a: 196)

Like a number of feminist scholars whose research projects have stemmed from an attempt to come to terms with dynamics affecting their own lives (see Kobayashi 1994; Moss 1994), my research into temping had its roots in my own experience working as a temp while I was an undergraduate in England.[18] Given an unshackled notion of the field, this very experience effectively served as my first round of fieldwork in the industry. 'Experience', though, is an easily abused category, and I do not mean to infer here that my work as a temp went on to guarantee an instant rapport with the temps I interviewed in my fieldwork in Vancouver. Turning experience into an origin for ethnographic authority in such ways leads directly to the dangers of identitarian absolutism critiqued by Kobayashi (1994).[19] Experience is inadequate as an *origin* of explanation, and, instead, as Joan Scott has suggested, it is more usefully understood as 'that which we want to explain' (1992: 38). It was in just this sense, then, that I began the research on the temping industry in order to come to terms with my own earlier experiences. Ironically, however, in doing the fieldwork, I found myself involved in new experiences that sometimes repeated some of the same contradictions that had shaped my time as a temp. Not least of these was my position as a man – first as a worker and then as a fieldworker – in an industry that predominantly employs women. It is to these contradictions that I would next like to turn after a few comments about the limits of academic male reflexivity.

The problem of male self-reflexivity

This is already as McDowell (1992: 56) notes, 'a profoundly self-reflexive moment in human geography', and perhaps it is more productive now to distinguish between different types of self-reflexivity. Although my aim in what follows is to follow the example of self-situating commentary presented by the 'Women in the Field' writers, my position as a man makes a difference. When men turn self-reflexive problems arise. As a form of gender-alert self-examination such work may halt men's abdication from what Elspeth Probyn (1993: 47) calls 'the responsibility of speaking their own bodies', and, for this reason, it may also thwart the tendentious but common practice of associating 'gender issues' solely with women. But, as Probyn also indicates, the straight, white and propertied male voice has for a long time been the only voice allowed to wax autobiographical – even if it did so transcendentally. Contemporary attempts at self-reflexivity from such privileged positions, however well-intentioned, still carry with them something of this history of hegemony, and one of the results tends to be a moralizing mixture of introverted angst and anger. '[This] self-critical mode of reflection', says Julia Emberly, often manifests the worst attributes of the Christian moral imagination. A religious rhetoric of fear, guilt, redemption and absolution emerges. A narcissistic return to self-centering lurks on the horizon. The male investigating subject, having been confronted with feminist politics, sees himself as a sacrificial son, a symbol of atonement for the original sin of patriarchy, the rule of the father.

(1993: 85)

I find this critique particularly pertinent to the positions of men like myself whose experiences have been basically heterosexual. As I proceed in what follows to try and turn my position as a 'participant-observer' into that of an 'observed participant', I want to avoid the confessional logic Emberly highlights as much as possible. In particular, it seems vital to remember that redemption and absolution do not lie around the corner, and that, instead, auto-critique needs to be persistent, constantly problematizing the moments in which interpretations others might make are marginalized through a return to maudlin, men's-movement type self-centring.[20]

Between fieldwork on temping and temping as fieldwork

Some of the experiences I had as a temp had less to do with gender and were more related to economic processes. Although, of course, class and gender remain inextricably related, some of the situations I faced – such as the struggles over pay-rates, and my inability to get assignments because 'things were slow' after the 1987 Stock Market crash – served better as indicators of class processes, and, as such, when suitably contextualized, crystallized a broader pattern of economic vulnerability (see Sparke 1994c). By contrast, my more obvious experiences of gendering in the industry called for explanations that could come to terms with what, paradoxically, they did *not* embody. Being a man makes a significant difference in temping. For one thing, I was commonly sent out on unconventional, masculine-coded assignments, like the jobs I did driving a bulldozer and a fork-lift truck. But more than this, even when I was doing secretarial, filing and telephone-answering work, I had experiences which, if they illustrated anything, it was more as exceptions that proved the rule: in this case, the rule of the patriarchal relations that feminize temping as trivial but necessary office house-keeping.[21] Unlike many temps, I was not so quickly marginalized because as a man and a student I had what was treated as extra curiosity value. Likewise, when I answered the telephone with a male voice callers often took me for the manager. Add to these assumptive dynamics the fact that as a young single student I did not experience the same heavy economic burdens that weigh down on the will to resist of many women who temp, and it all made for a different and privileged experience of life in the industry.

When it came to my fieldwork in Vancouver the same differences became more problematic. While I wanted to come to terms with the gendered dynamics of temping that had conditioned my own experience, and while my research was sensitized by a series of questions stemming from feminist scholarship, I was still basically positioned as a man 'going into the field' to interview women. However much I sought to eschew the traditional enframing of the field, I was still in a situation where it was difficult to articulate a space of between-ness in which my interviews might be experienced as something more useful than the customary mix of invasion and appropriation. I had wanted, for example, to follow the format of long personal interviews that had characterized Rosemary Pringle's (1988) tremendous study of and with secretaries in Australia. However, as I introduced similar questions about feelings and the gendering of work in my own interviews, how were the temps I was interviewing to know that my project was not motivated instead by a prurient, peep-show type patriarchy?

Even before the interviews began the way in which my research was implicated in the male researcher/female subject structure had effects. When I rang up someone who agreed to do an interview we had to work out a place and a time that would be convenient. Sometimes this led to talk about good and bad cafes and malls, or about the length of lunch hours and the working day. Subsequently, we usually described what we would be wearing or carrying. Especially with younger women who did not know me, this set of exchanges sometimes made *me* feel that it was all a little like a blind date, and I definitely did not want the women themselves to be obliged to negotiate that implication. However, my efforts to allay potential fears of this kind were themselves not unproblematic, and my occasional attempts to say something like, 'I know this sounds a bit like a date, but really, please don't think that's what I'm trying to set up', ironically mimicked, however awkwardly, the patronizing remarks temps often hear as secretaries from bosses. Both circumstances involve a denial of women's abilities to make their own interpretations and decisions.

In another instance, the questions of gendered research became still more explicit during the course of an interview with a young woman whom I will call Karen Edwards. The interview was in the coffee-bar at the Vancouver Art Gallery on a Saturday lunch-time and did not begin before a rather confusing 10 minutes of walking past one another. Karen's sister had also come along, and so, having sat down opposite both of them, I began to ask some preliminary questions.

Matthew:	So how long have you actually temped now?
Karen:	Since the beginning of January – so about three months.
Karen's sister:	Can I just ask a question?
Matthew:	Yeah, sure.
Karen's sister:	Why . . . – uh well there's nothing wrong with this, but couldn't you guys have done this on the telephone?
Matthew:	Well, yes and no. I mean, like yes we could have done, but later on I want to talk more about feelings, like about how one feels about work, the way you're treated and positioned by people and stuff like that.

Karen:	So you need to get facial gestures and things.
Matthew:	Well yeah – and like on the phone, I've tried it and it's really weird – people give you a quick answer like 'No that never happened', or, 'Yes it's OK.'

(Laughing)

Karen's sister:	Just wondering.
Matthew:	So yeah, it's not my way of trying to get dates, right – that's not what I'm doing.

(More laughing)

Karen's sister:	I didn't want to say it like that. It just made me, well [laughing] think.
Karen:	Well it's a good way to meet people in a new country I guess.
Matthew:	Yeah right . . . So – anyway – what kinds of jobs have you actually been doing on assignments?

It seemed to me that Karen's sister initially felt that the interview, with its awkward beginning and simple start-off questions, looked like a rather duplicitous attempt on my part to arrange something like – 'there's nothing wrong with this' – a date. She told me that she had come along because they were both going shopping across the road at Eatons, but I think the reasons why an older woman might accompany her younger sister to an 'interview' with an anonymous man also relate directly to very real fears about safety. Karen, herself, with a number of other phrases like 'so you need to get facial gestures', seemed to have wanted to preserve the idea that the interview was properly 'academic'. Yet, she was not at all perturbed by the possibility of its ulterior function as a way of meeting people. Indeed, putting her two ways of articulating my position together appeared – *I think* – to recreate quite well a picture of me as an academic male tourist: a late twentieth-century version of the Englishman abroad on his grand tour of educational improvement.

Another problematic dynamic exemplified by the interview with Karen, and, indeed, by the preceding discussion of it, concerns the politics of interpretation. It was, after all, my own reading from my own position that led me to bring up the dating scenario in the first place. Likewise, I have here only presented my own view – another objectifying verb for knowledge – of what I thought Karen and her sister were thinking. One further way of illustrating how my own position was shaping my interpretations became obvious in earlier versions of this chapter itself. In it I had discussed the laughing that was part of the exchange, and had commented on how I thought my two interlocutors might have been thinking that the whole event was a bit of joke. However, as one of my feminist colleagues pointed out, this was a notably masculine reading of the laughing. Instead, she suggested that, from her perspective, it could equally be read as the laughing of women dealing with feelings of anxiety. Clearly the differences in these interpretations highlight the specificity of my own reading. As such they also point up the more general possibilities of exclusion and violence that my own representations of my research – like those of any other scholar – impose on the researched.

The dynamics of gender, or at least the more awkward, problematic and potentially oppressive aspects of gender relations, did not always become so thematic in my fieldwork. When I was talking to older women, for example, my 'Englishman abroad' position was subject to quite different readings and feelings. Apart from talking about the poor availability of Marmite in Vancouver or decent coffee in England, we often compared notes on the differences between the organization of temping in Canada and the UK. Not only was this valuable for my research – introducing such issues as overtime rules and the way agencies dodge providing holiday pay by cancelling assignments over national holidays – it also brought attention to common class experiences, shared by us both as men and women workers in workplaces globally homogenized through international capitalism. In other words, it exemplified for me what Haraway (1991a) describes as 'the unexpected openings situated knowledges make possible'.

Perhaps the most unexpected of openings and the most developed articulation of between-ness in my fieldwork overall, however, was the way in which temps themselves described how they felt that they engaged in a form of fieldwork *as* temps. Constantly tracking back and forth from the rest of their lives to a job, and then to another job, and then another, temps move through the 'dialectic of experience and interpretation' that Clifford (1988: 34) describes as a defining feature of participant observation. However, unlike academic fieldworkers who assume a capacity to leave the field, temps' movements are ordered quite directly by capitalist economics. There is certainly little academic idealism driving this process along, no neo-colonial will-to-knowledge and no great sense of agency, but rather the much more practical need of working to earn money to live. As they move from office to office they cannot help but see and feel the differences and similarities. Moreover, temping is in this sense a form of participant observation with the emphasis heavily placed upon participation. There is none of the voyeuristic privilege that comes from the academic's material well-being resting on the 'interpretation' side of the dialectic. Temps have instead to participate with a will, they have to 'go native' in the new office as fast as possible. The following quotation from an older woman I interviewed was a typical ethnography of arrival, and a sensitive piece of fieldwork to boot.

> Normally I ask: How do you wish to have the phone answered? Who have I got? Who are the names and where are the numbers? And normally it's: Oh no, sorry we don't have a list, or so and so do you have a list, and so and so never does. And usually the desk, if you get a desk, isn't stocked, there's no paper, you have to ask. I've never been told, well here is the copy room, here is the fax room, here is the mail room. All it would need is a little map or something. And this is what happens. And you have to find it all out yourself. And if you ask any questions, you're no damn good. You've got four hours to make your mark before you're pulled off or left on the job. You ask too many questions it shows you're incompetent.

Going through such arrival routines on a regular basis, temps experience the power relations of the hegemonic and routine, and in doing so they become knowledgeable about hegemony. The possibility of such knowledge production may even be read into the sorts of facile encouragement dished out in agency magazines. For instance, Kelly's Workstyle$_{TM}$ pamphlet, offering 'tips from Kelly Services® for managing your work and life style', used as a quote of the month an insight of Eleanor Roosevelt's: 'You gain strength, courage and confidence by every experience in which you really stop to look fear in the face.' It should be emphasized, though, that such fear can be very real for temps, and, unlike the academic fieldworker who might, like Geertz (1973: 7), be urbanely exercised by the problem of distinguishing wink from faked wink, temps are obliged to interpret such cultural distinctions in a context where they are far more closely felt and sometimes quite sexually threatening. One interviewee assessed a typical patriarchal pattern as follows.

First day in the office all the men come by with some precarious question. There's often this really silly need for them to comment on the way you look – 'You look very professional', 'You look very nice' – and you have to get through all this stuff and still look professional. But I think the problem is not so much the compliment but the system of expectations you have to work out quickly.

There were many similar observations I recorded during my fieldwork, but the point I hope is already clear: temps too are agents of knowledge and interpretation. Far from being stationary others, their movements as supplementary workers afford them a position as producers of supplementary and, as such, potentially disruptive information. They see how the conventional is organized, varyingly and yet repeatedly, and in doing so they also see that it is not universal, something that is produced and which need not therefore be necessarily taken for granted. These are some of the classic characteristics of doing comparative fieldwork. But the differences distinguishing temping from professional academic fieldwork also need to be noted. Not only do temps experience the strangeness of the field in more oppressive ways, they also have few of the resources granted academics. As was noted by the temp commenting on arrival routines, they are rarely given a map before they set out, and the chance of meeting an informant tends to be foreclosed by the routinized discipline of most modern offices.[22] Moreover, temps who don't go on to become academic fieldworkers are rarely able to have their ethnographic observations listened to, let alone read in scholarly texts. Indeed, for them accruing the knowledge without anyone particularly wanting to hear it can become a practical problem. 'It's one of the reasons I'm glad I'm not a temp anymore', said one of my interviewees.

You always have to forget. Like when you get to an office and have to learn everything in 10 minutes. I filled up my head with so much garbage that way, that I had to train myself to forget everything. And then, of course, when I got good at forgetting, I got those dumb recalls asking for me to go back because 'she knows our routine'. And then, of course, by that stage I'd normally got it right out of my mind.

It was in the hope of listening and giving voice to people who are in this way obliged to learn only to forget and then learn again that I set out to record their observations and feelings. This led to some very detailed accounts of the industry which I found invaluable in my own research into its gendering and political-economy. Moreover, like England (1994: 85) in her earlier work, I was told by many of those that I interviewed 'that they found the exercise quite cathartic', allowing them to verbalize things that no one else had ever really wanted to hear. At same time, however, and also like England, I was mindful of how this did not prevent my interviews from also becoming moments of misappropriation, appropriation turned politically damaging, even oppressive. Particularly in exchanges like the one with Karen and her sister, I became aware of how my argument about giving back through listening could become little more than a rationalization for a patronizing form of business as usual. For this reason, I fully concur with feminist critics such as Gilbert (1994) who argue that attempts to meaningfully articulate and extend a space of between-ness are severely limited.

A CONCLUSION

[F]ield and home are dependent, not mutually exclusive, terms, and . . . the lines between fieldwork and homework are not always distinct. . . . Home once interrogated is a place we have never before been.

(Visweswaran 1994: 113)

Against the logic of the confession, I do not want to end on a note of self-critical despair. While some of the masculinist framings of the field are not easily displaced, their renegotiation remains a possibility even for men. At a practical level, for example, I would another time organize my interviews quite differently by enlisting the assistance (paid assistance preferably) of women colleagues. Even if it was just in the setting up of interviews, such help might allay the potential fears of would-be interviewees. Similarly, I would also in the future want to organize a more collaborative, focus group form of research that could also serve as the basis for getting temps together. More generally, though, it seems sanguine, even arrogant, to hope for immediate political and organizational advances through such refashioned fieldwork alone. Such aspirations would appear, in fact, to begin to forget how the notion of between-ness opens the possibility of multiple spaces for social change, some of which might be discontinuous from the research if not from the arguments to which its findings can contribute. They thus risk absolutizing the so-called space of between-ness as some form of reified 'Third Space', turning it too into a fixed and fetishized foundation that simply consolidates academic authority through another anemic geography. For the same reason it seems critical to heed Visweswaran's (1994: 102) warning that: '[r]ecent proclamations that "the field is everywhere," even when coupled with critiques of fieldwork, do little to unsettle the epistemological weight fieldwork signifies [within academic disciplines].' It is vital, therefore, to underline how scholars like Katz began their rearticulation of between-ness, not just by linking it with reflexivity about the situations of academics, but also with the call for 'strategic *displacements* that merge our scholarship with a clear politics that works against the forces of oppression' (1994: 67). As Damaris Rose notes, '[i]t is important not to lose sight of this activist goal as the study of "gender issues" becomes more accepted into the mainstream of academic disciplines such as geography' (Rose 1993: 58). So far, though, my own direct activism around temping has been limited to a few gestures of solidarity with temps and a public employees union.

Here, nonetheless, is a point where a dynamic understanding of the space of between-ness also points beyond the crippling either/or of guilt or revolution. Remembering our always already embeddedness in between multiple overlapping fields can also function to *continually* remind us of how academic freedom brings with it what Spivak (1992: 7) calls 'the freedom to acknowledge insertion into responsibility'. Located between fieldwork and the academy we are persistently obliged to acknowledge such responsibility in freedom. It is for this reason a responsibility that will not go away. It does not promise absolution, it has no fully redemptive end in sight, and it demands only more reflexivity. But in Spivak's feminist, decolonialist and Marxist sense, it also urges us to continue to search for the emancipatory possibilities implicit in a geography where as Sauer himself once said '[a]cademic freedom must always be won anew' (1956: 299).

ACKNOWLEDGEMENTS

I wish to thank Nancy Duncan, Jennifer Hyndman, Cindi Katz, Vicky Lawson, Debbie Leslie, Donald Moore, Pam Moss, Heidi Nast, Anila Srivastava and Gillian Rose for their comments on a variety of earlier versions of this chapter.

NOTES

1 I am here attempting to highlight what I feel is a quite practical implication of Derrida's argument about the disseminatory power of repetition turned displacement in *Dissemination* (1981). Fieldwork passed down as method by the great father figures of geography can be seen in this more arcane register as open to *re*-working by their undutiful daughters and sons. However, and contrary to a universalizing 'Law of the Father' such as Sauer's, a deconstructive reading would suggest that this can never be secured by edict. It becomes possible, only as a 'semination that is not *in*semination but *dis*semination, seed spilled in vain, an emission that cannot return to its origin in the father', Spivak, (1976: lxv). For a frank discussion of the performativity of gender itself see Butler (1994) and for some reflexive geographical performances see Bell *et al.* (1994).

2 Heidi Nast and Virginia Blum (under review) argue that Katz and Smith confuse Henri Lefebvre's terminology here. While their argument about the dangers of spatial metaphorization turns to the French philosopher's historicization of the modern emergence of a reified and dehistoricized conception of space, they replace

his name for it – 'abstract space' – with another name – 'absolute space' – that Lefebvre himself connects to *pre*-modern and, historically 'absolutist' conceptions of spatiality. See Lefebvre (1991) especially Chapter 4, 'From absolute space to abstract space', and the helpful description of the distinction by Stewart (1995). Lefebvre's distinction noted however, I still think that the English name 'absolute space' better evokes the notions of desacralized but containerized and emptyable space that Katz and Smith use it to describe. 'Abstract space', by contrast, is so full of implications that it invites only further confusion.

3 The possibilities presented by Harding's (1991) argument about contradictory positionings are obviously extensive. Here, in the context of this volume, I concentrate on the question of gendered positioning in particular. This clearly downplays other questions – for example, about racialized and colonial positions in fieldwork – but not, I hope, in a way that presents them as irrelevant. Instead, I consider them as future work.

4 I return to the masculinity of objectification in a later part of the chapter. However, for a more sustained critique see Haraway's ovular essay, 'Situated knowledges: the science question in feminism and the privilege of partial perspective', (1991a).

5 I am referring here to the distinction Rose draws between the 'hegemonic', scientized masculinity she illustrates with Hagerstrand's 'time geography', and the 'aesthetic masculinity' she finds in humanistic and cultural geography.

6 The reference to *fort/da* is to the self-defining and arguably masculinist space game played with a reel by Freud's young grandson Ernst. It is discussed in Freud (1959). I have addressed the masculinity of the game and the questions it raises at length in Sparke 1994a.

7 Clifford Geertz notes as much – however serenely – when he distinguishes his vision of scholarship from tourism. 'In itself', he notes, 'Being There is a post-card experience ("I've been to Katmandu – have you?"). It is Being Here, a scholar among scholars, that gets your anthropology read [. . .] published, reviewed, cited, taught' (1988: 130). However, the anthropologist Orin Starn has more critically highlighted how 'as a reinscription of the imagery of separation and stability, the metaphors of "home" and "field" may be counterproductive in the development of languages and frameworks that reckon with what David Harvey calls the "time–space compression" of the contemporary world' (1994: 35).

8 As Rose (1993) suggests, this is something that Katz and Smith begin to do themselves when they lay claim to a specific concept of real space as the ground for spatial metaphors, see 'As if the mirrors had bled: masculine dwelling, masculinist theory and feminist masquerade' (this volume). In such claims there remains, of course, the danger of neglecting equally 'real' and material spaces as they are imagined and experienced from other perspectives. Nevertheless, I also read in Katz and Smith's problematization of reified spatial metaphors a gesture of critique that invites an openness to precisely such other perspectives.

9 LaCapra, 'Who rules metaphor?'. This was a review of the logocentric impulse in Paul Ricouer's, *The Rule of Metaphor: Multidisciplinary Studies in the Creation of Meaning in Language* (1979).

10 She notes that

> masculine or phallocentric discourses and knowledges rely on images, metaphors and figures of women and femininity to support and justify their speculations. [There is a] disavowed corporeal and psychic dependence of the masculine, with its necessary foundations in women's bodies, on female corporeality it cannot claim as its own territory (the maternal body).
>
> (1990: 74)

11 Those who seek to paint over the violence of fieldwork by concentrating on its aesthetic dimensions would do well to read Haraway's account of how the camera became privileged over the gun within the context of a scopophilic law of the father. 'The true father of the game loves nature with the camera; it takes twice the

man, and the children are in his perfect image. The eye is infinitely more potent than the gun. Both put a woman to shame – reproductively' (1989: 43).

12 Derrida speaks more generally of how 'metaphysics has erased within itself the fabulous scene that has produced it, the scene that nevertheless remains active and stirring . . . an invisible design covered over in the palimpsest' (1982: 213).

13 Although, as Pam Moss has pointed out to me in private communication, saying that blasphemy comprises dissent without apostasy, may not adequately reflect the serious and, in some ways, faithful forms of study conducted by feminists that remain committed to challenging the status quo.

14 Pred also critically documents here what I would call Carl Sauer's romanticized white supremacism. Such criticism would no doubt also seem to threaten the imperialism of the sacred fire.

15 This put down seems to me to be bolder than that of George Marcus who, as an anthropologist, recently mocked attempts at ethnographic reflexivity by geographers, dubbing them '"[m]ore (critically) reflexive than thou"', (1992). While there was critical blasphemy that perhaps discomforted Marcus in the subsequent essays of Crang, Katz, Keith and Rogers, his commentator's worry that reflexivity can become 'the mode of a rather puritanical, competitive assessment among scholars' (1992: 489), did nevertheless suggest a danger which I feel is serious, and which I discuss in relation to my own masculinity below.

16 She notes in conclusion that

> [a] written text is merely a point amidst a continuous fabric of other texts that includes all communicative forms through which researcher, researched and institutional frameworks are relationally defined. Such contextualizations are essential if we are to carry out the kind of collaborative, global and otherwise transgressive kinds of research that presently peppers feminist geographers' horizons.
>
> (Nast 1994: 62).

For a critique of how textualism has been limited so as to exclude feminist work in anthropology see Deborah Gordon's critique of the role of 'Writing' (1988: 7–24) and, the further contextualization of its racism written by bell hooks (1990: 123–33).

17 See also Teresa de Lauretis's post-Althusserian description of how critics conducting feminist critique lie 'both IN and OUT side ideology' (1987: 10).

18 The name 'temp' is the popular abbreviation for temporary workers whose largest contractor in North America, the misnamed Manpower Services Inc., now employs more people annually than General Motors. For a discussion of the industry's political economy see Sparke (1994c). For another series of quasi-ethnographic critiques of temping see the testimonials now being printed in the popular zine *Temp Slave* from Keffo, POB 5184, Bethlehem, PA 18015.

19 See also James Clifford's discussion of 'fables of rapport' used to 'narrate the attainment of participant-observer status' and thereby establish 'a presumption of connectedness, which permits the writer to function in his subsequent analyses as exegete and spokesman' (1988: 40).

20 A major problem with a moralistic approach, of course, is that along with self-centring, the marginalized get marginalized still more. It leads to a dead end. Nast puts it like this: 'Guilt that centers merely on the existence of this inequality and not on how the inequality can be transformed is therefore unproductively paralyzing' (1994: 58).

21 I discuss the patriarchal relations structuring temping at length in Sparke (forthcoming).

22 I should note, though, that some of these deficiencies have now begun to be addressed as a matter of capitalist expediency. 'Provide some information on the "culture" and "norms" of your organization so that temps can fit in comfortably' advises *The Office*, 111, 1990: 59–60. In a similar vein *Supervisory Management* for

August 1989 recommends the following: 'When the temporary employee arrives, give a brief tour around the office. Make sure to show the person where to put his or her coat, the locations of the rest room, the water fountains, cafeteria and so forth. Also include in the tour the supply room, the copy machine, and any other equipment the temp will need to use', (pp. 26–7).

REFLECTIONS ON POSTMODERN FEMINIST SOCIAL RESEARCH

J.K. Gibson-Graham

Until quite recently feminist empiricism and feminist standpoint theory have offered epistemological positions that have been the basis for a phenomenal growth in feminist social science (Harding 1986). Empirical studies conducted from a range of theoretical perspectives (radical, socialist and liberal feminist) have all in some way affirmed the existence of women's experience as a source of privileged understandings, if not the basis of an alternative social science. Now, however, the deconstruction of 'women' is having profoundly destabilizing effects upon feminist theorizing and research (Barrett 1991).[1]

Wendy Brown writes of the 'palpable feminist panic' that has arisen as the situated and subjective knowledge of 'women', gleaned, for example, from ethnography, oral history material or consciousness-raising groups, has come under attack for its presumption of representing the 'hidden truth' of women or women's experience.[2] While the turn to postmodernism has engendered a plethora of exciting philosophical, political and cultural endeavours that tackle the essentialism around women embedded in both feminist and non-feminist texts,[3] feminist social analysts find themselves confronting an ironic impasse as the unifying objects of our research dissolve before our eyes.

This chapter tells the tale of a social research process which has been shaped by the flux of current feminist debates. It takes up some of the problems of 'doing gender' outlined by other feminist geographers such as McDowell (1992), Dyck (1993), Rose (1993) and Pratt (1993). Like them, I take seriously the challenges posed by postmodern theory to feminist social scientific research. If we are to accept that there is no unity, centre or actuality to discover for women, what is feminist research about? How can we speak of our experiences as women? Can we still use women's experiences as resources for social analysis? Is it still possible to do research *for* women? How can we negotiate the multiple and decentred identities of women? In this chapter I try to reflect self-consciously upon these questions as I discuss the research methods employed in my own project.[4]

MINING AND REPRESENTATION

During 1987 and 1990 when I was researching the development of new mining localities in Central Queensland, I documented stories about the activities of women along with those told by company managers, union representatives and community workers. I soon noticed that two 'identities' or 'constructions' were available in relation to the women I was concerned with. One was the representation purveyed by mainstream Australian social analysts and service providers influenced by the liberal community studies tradition. In this discourse 'mining town women' are a client group who have needs for better social, psychological and health services. They are portrayed as independent and self-reliant but defensive, vulnerable, cautious of emotional commitment, lonely, isolated, stressed and traditional.

The other representation was that purveyed by socialist and socialist feminist analysts in Australia, the UK and North America. In this international discourse 'miners' wives' are situated as auxiliary members of the industrial proletariat, the feminine face of the solidary working-class mining community that holds such a hallowed place in left sociology. While researchers studying the newer mining communities of Australia and North America have emphasized the differences between women in these more affluent and isolated new towns and those in traditional mining communities of the British, Welsh and Appalachian coalfields, miners' wives are still constituted within a discourse that foregrounds the functional overlap of capitalist exploitation and patriarchal oppression. Proletarian first and women second, miners' wives are subsumed to the fictional identity 'working class' and relegated to the status of Other within this totalizing conception.[5]

The focus in the socialist feminist literature upon industrial disputes, in which women are expected to express their real identities through solidarity with working-class men, situates this literature solidly within essentialist Marxism-feminism where consciousness is true or false and subjectivity structurally constructed and constrained. Most of the studies within this tradition highlight those occasions when miners' wives invert their Other status, come to recognize their true class alignment, and heroically join, lead or hold their men to authentic working-class consciousness and action.

As I collected stories of women's political involvements in mining towns, I began to realize that research could not 'add women in' to the picture without situating them with respect to one or the other representation.[6] Yet these characterizations seemed to construct narrow, unidimensional identities and subject positions with negative and disciplinary overtones. Both the client/victim/pathologized individual representation of liberal discourse and the proletarian/militant/supporter cum leader representation of socialist discourse denied the potential for a multiplicity of political subjectivities to emerge. And both seemed to actively organize women out of any independent involvement in either industrial or gender politics.

The stories I collected had a complex and somewhat ambiguous relationship to existing discourses and subject positions available to women in coal-mining towns. Some of them undermined the representation of 'mining town women' as individualistic and depressed, unwilling to connect with other women or contribute to community activities. But the traditionalism of the gender relations in these stories reinforced mainstream representations of women as personally and politically dependent upon men. The stories also overlapped with the socialist feminist discourse on 'miners' wives' as members of the working class, but at the same time allowed glimpses of other processes by which different (non-working-class) subjectivities were continually being crafted, and sometimes enunciated in action (Kondo 1990). Recognizing the identities 'mining town women' and 'miners' wives' had allowed me to see both as regulatory fictions masquerading as self-evident categories of analysis, each of which positioned women in mining towns in subjugated positions.[7]

This initial process of deconstructing the categories 'mining town woman' and 'miner's wife' had enabled me to identify their politically powerful disciplinary and exclusionary effects. At the same time I had begun to see glimpses of alternative subject positions and political identities for women in mining towns with which I could interact. It occurred to me that producing alternative discourses of gender and mining town life was one way of liberating alternative subjectivities for mining town women.

The question that soon emerged was, why create an alternative discourse, an alternative voicing, and where was its audience? Who was interested in new subject positions for women in mining towns? And why construct alternative subjectivities for mining town women if 'women' in general had disappeared? The usual answer to such questions harks back to the political project of feminism and the central role that research plays in the 'liberation' of women. But the political project of postmodern feminism is now a matter of considerable debate.

RESEARCH AS POLITICS/POLITICS OF RESEARCH

In dissolving the presumed unity of women's identity postmodern feminism has liberated knowledges and given rise to fruitful theoretical controversies as to who women 'are' and how to 'know' them. At the same time, however, Brown's 'palpable feminist panic' (mentioned at the outset of this chapter) seems to have migrated from the realm of theory into the political realm, where the identity 'woman' has usually been constituted as the *necessary* ground of feminist political action. Feminists have historically claimed that as 'women' we are dominated and oppressed, and feminist politics has staked its legitimacy upon the assumption of this shared or common, but importantly, *subordinated* identity. To surrender epistemological claims about women's shared identity has signified, for many, giving up the structural and moral position from which to organize politically to overcome oppression (Brown 1991: 75). Without unity of women's identity, many critics see postmodern feminism as opening the doors to fragmentation, factionalism and political disempowerment.[8]

It seems that, for many, a paradox has emerged – as knowledge has been liberated, politics has been shackled. While feminists may agree that in theory, difference empowers, when it comes to politics many still hold to the adage that 'united we stand, divided we fall'. Ferguson (1993) argues that feminists must accept this contradiction and learn to live with the inevitable tension between articulating 'women's experience' and deconstructing the texts that represent and enforce this presumed commonality. Rather than insisting on a real, originary essence that defines all women, Ferguson advocates constant movement between the (strategically essentialized) representation of women's experience and the (strategically non-essentialized) deconstructivist practice of undermining fixed categories of identity and gender. In support of this view, Pratt (1993) calls for an 'equal commitment' and 'continuing dialogue' between these two moments of research practice. Barrett similarly positions the two moments in opposition: 'So it is an issue of whether one wants, speaking as a feminist, to deconstruct or to inhabit the category of "women"' (1991: 166).

I see a danger in posing these moments as opposing practices in irreducible/ ironic tension, the one associated with grounded commitment, the other with relativism and disaffiliation. Surely all deconstruction, or the tracing of 'how we produce truths' (Spivak 1989: 214), is done from a specific theoretical and political entry point from which further interpretation also proceeds. Seeing one posture as less political because it highlights difference, and the other as more political because it highlights collective identity, seems to suggest that the politics of identity is the only viable political form (for feminists at least). It also implies residual loyalty to the modernist separation of theory and practice – that conception of a knowledge/theory existing separate from and prior to change/politics (we understand the world in order to change it). Practice or politics, in this formulation, can only be enacted by a collectivity of subjects all identically positioned vis-à-vis the structure of power that has been rendered visible by theory. What this conception betrays is an interesting failure to see knowledge and its production as an *always already* political process.

It seems that what is needed is a rethinking of the relationship between politics and research. Following Foucault, I would see postmodern feminist politics starting from the assumption that power is everywhere inscribed, in and by women, as well as by men, in theory as well as in practice, in difference as well as in unity. Thus the process of theoretical production is as much a political intervention in changing power relations as is self-consciously (identity-based) political organization. There is no prior reality or unified identity to be accessed or created by research from which we can launch a programme of change. There are, however, existing discourses that position subjects in relations of empowerment and disempowerment. The ways in which theory and research interact with these discourses have concrete political effects.

As a social researcher, I was interested in the circulation of alternative discourses on women in mining towns within communities of interest not readily touched by academic writing. One mode of circulation open to me was the research process itself. Encouraged by feminists such as Brown and Weedon to engage in conversation and public discourse,[9] I wanted to move beyond a purely literary discursive intervention and into 'the field'.[10] I embarked upon a research process that attempted to involve women as 'knowing subjects' in the 'always already' political nature of the research process.

Immediately I was confronted by the practical dilemmas of how to include women in mining towns in the process of discursive deconstruction and the circulation of new discourses. One alternative was to attempt direct intervention in power relations between men and women by embarking on a project of action and participatory research. But this method of research relies upon an identification between researcher and researched and the discovery of a shared subject position from which political intervention can be discussed and enacted (Reinharz 1992). Without an assumed basis of unity between women could these research methods still be employed? In pursuing my idea of social research as a public engagement in the construction of alternative discourses, I was forced to rethink methods of action research in terms of postmodern feminist social research practice. The last section of this chapter tells the story of the research project that developed when an industry restructuring initiative instituted a shiftwork schedule called 'the 7-day roster'. This very demanding schedule of shifts was justified by the miners' union in terms of its negligible physiological effects on the individual miner and its positive impact on the miner's total annual income; but it proved to be extremely disruptive to family and community life.

IDENTITY, DIFFERENCE AND POSTMODERN FEMINIST ACTION RESEARCH

As my own research and that of others had established, in mining towns women are marginalized by many processes (Gibson 1992a; Sturmey 1989). I decided to employ a number of miners' wives as co-researchers in the project in an attempt to confront some aspects of women's marginalization.[11] The women employed had to be experiencing life with a shiftworker on a 7-day roster. They also were selected on the basis of their stage in the life cycle and family formation. The twelve participants (three each from four different mining towns) were actively involved in the research design and questionnaire formulation and were trained as interviewers at an initial two-day workshop to conduct six recorded in-depth discussions with their friends and acquaintances.[12] In this way it was hoped that an established rapport would exist between interviewer and respondent and would be the basis for a more relaxed and revealing interview experience. The women received payment for their interviewing work and all expenses were covered to allow them to attend two workshops at a location that was central to all the towns. At the second workshop held later in the year, preliminary results were analysed, qualitative results discussed and possible interventions outlined.

I saw this project as a modified or postmodern form of action research. While I actively involved women in the process of researching their own situations with respect to shiftworking partners, the project had no underlying agenda of consciousness raising and direct group action. The initial training workshop and the later feedback workshop incorporated discussions of consciousness and action but there was no expectation on my part or that of the participants that a feminist political programme would or should emerge. Instead, in the workshops and over the kitchen tables where one-on-one interviews were conducted, the research project created and cultivated spaces in which a feminist politics (the transformation of gendered power relations) was performed in conversation and group discussion.

In the process of creating this political space, 'place' assumed some import-ance. Unlike in cities where people 'live together in relations of mediation among strangers with whom they are not in community' (Young 1990: 303), in small mining 'communities' people foster links with each other through 'being the same' and excluding and ostracizing anyone who is different. Removing women from their 'communities' was the first step in creating a discursive space in which to construct new political subjectivities.

Within the research team, various other barriers stood in the way of political conversations. Many of the differences which divide and structure our everyday social experience were present in the group that assembled at the first training workshop – urban–rural born, married–unmarried, educated–uneducated, older–younger, childless–mother, adopted–not adopted, wealthy–poor, traditional–feminist, spiritual–materialist, new age–mainstream, fat–thin, forthcoming–shy, amongst others. In this group the decentredness of women's 'collective identity' and the overdetermined nature of subjectivity were patently obvious.

Throughout the workshops members of the research team inadvertently explored their differences with each other and with the other women they had interviewed, and with Joanne (the freelance community worker I employed as a co-researcher and local facilitator) and myself. In this space away from family, friends and 'community' women felt liberated to air differences without forcing conformity. I heard many comments prefaced by 'My life's not like that . . . my husband's not like that . . . of course things were different in those days than now . . . well, you're younger than me . . . some women must live in a very different situation than I do.'

At the outset of the research, for many of the participants the lack of identification with each other was uppermost in their minds. But by the end of the second workshop both the idea and the act of 'partial identification' had become more developed.[13] Interestingly, what emerged was not what I would see as identification around the shared experience of women (the recognition that *as women* we shared a common 'problem'). In fact personal differences in gender experience widened on many fronts. What took place was identi-fication with respect to common problems of a very specific kind (ones that many women would not share) – living with a shiftworker (or in the case of Joanne and myself, living with a self-employed partner who worked long and irregular hours, often including weekends); particular place-specific forms of male discipline; union, company and university reluctance to consider family life in industrial relations.

In a sense my research process was constructing a partial but shared, externally related identity, and beginning to create a public knowledge about mine shiftwork and family life, about terror in the face of male power over women's ability to speak out, about women's mistrust of each other. The fiction of the 'mining shiftworker's wife' I was encouraging or imposing became a momentary reality – a basis for communication about many of the contours of power affecting political activism in mining towns. This comment was made with acceptance and resignation:

We're powerless in the face of decisions about the roster – the men won't listen to just us.

Other comments were made with surprise and consternation:

Many women confessed to hating the 7-day roster but they refused to be interviewed. Their husbands were forbidding them to be involved.

Even women who are normally very strong and stroppy said they couldn't do it.

The men are so suspicious, so them and us. They thought this survey would be used against them.

In the process of research the participants became open to otherness and aware of their own political capacities:

This has made me realize that not all people can cope with the roster system, and just because things in my part of the world run smoothly, does not mean there is nothing wrong with the town I live in.

The research helped me rationalize my thoughts about the 7-day roster and was very helpful in coming to terms with many personal issues. It boosted my self-worth and gave me a sense of achievement and involvement in community and value other than as a housewife/mother.

In the political space created by this research project a new discourse of mine shiftwork and a new subjectivity of the 'mining shiftworker's wife' started to emerge. As women engaged in the myriad conversations that formed part of the research, they actively displaced the existing discourses of 'mining town women' and 'miner's wife' that confined their subjectivities. Out of this process a new subject position has developed – one that is focused on the gender division of labour and the impact of industrial conditions and disputes on relations within the home.

As companies discuss the possibility of introducing 12-hour shifts in the mining industry and long-distance commuting from the coast to new mines, the results of this research are circulating as an alternative way of thinking through the issues. A booklet that illustrates the research findings using cartoons and verbatim comments from women has been published by the miners' union and distributed throughout the region, significantly aiding the circulation of the discourse of shiftwork and family life.[14]

A new (the first) occasional childcare centre has been built in Moura after attempts by one of the research participants (who was motivated by the research experience to 'get off her bum') to set up a baby-sitting club failed. This woman took the research report along to the meeting to discuss the need for such a service and was able to influence the decision to establish the centre.

One of the women interviewed asked for the tape of her interview back, sat her husband down and made him listen to it. She then was able to broach her anger with him for spending all his days off from minework at the new farm they had just purchased. Their interviewer was pleased to report that 'now they're like a pair of newly-weds'.

At some level the research is challenging the established discourse of industry policy – its boundaries, the actors it legitimizes and its social effects. As women in all their diversity voice newly developed concerns around an issue of industry restructuring they enter, wittingly or unwittingly, an arena from which they have long been marginalized and excluded.

CONCLUSION

While I share no fundamental identity with any other person (as I am a unique ensemble of contradictory and shifting subjectivities) I am situated by one of the most powerful and pervasive discourses in social life (that of the binary hierarchy of gender) in a shared subject position with others who are identified, or identify themselves, as women. This subject position influences my entrée into social interactions and the ways I can speak, listen and be heard. In this sense I am enabled, as a woman, to research with other women the conditions of our discursive construction and its effects.

As a feminist researcher, I am coming to understand my political project as one of discursive destabilization. One of my goals is to undermine the hegemony of the binary gender discourse and to promote alternative subject positions for gendered subjects. I see my research as (participating in) creating identity/subjectivity, and in that process as constituting alternative sites of power and places of political intervention. Whether in conversation with mining town women or with other feminist academic researchers I understand my discursive interventions as constitutive rather than reflective, political as well as academic.

In my research I found the metaphors of 'conversation' and 'performance' much more useful in imagining a research strategy than the mining metaphors I had initially adopted. The mining metaphors constitute research as a process of discovery, of revelation: as researchers we reveal truths that are hidden from the untutored observer, contributing hitherto untapped resources to the permanent store of knowledge. By contrast, conversation and performance are metaphors of creation and interaction. Both processes are ephemeral, yet each may have long-lasting effects upon thought and action. Conversations can produce alternative discourses that entail new subject positions, supplementing or supplanting those that currently exist. These new subject positions crystallize power in new sites, enabling novel performances – individual or group interventions in a variety of social locations. In this way the creation of alternative discourses subverts the power of existing discourses and contributes to their destabilization.

This research process has provided insights for me into the practice of a new, postmodern feminist politics of difference. Action research need not focus upon the uncovering or construction of a unified consciousness upon which later interventions will be based. Action research can be a means by which we 'develop political conversation(s) among a complex and diverse "we"' (Brown 1991: 81). Within these conversations we create the discursive spaces in which new subjectivities can emerge. As the centred subject with its historic political mission departs the social stage, there is now room to talk of the inescapability of difference and the only/ever partial nature of identification. Yet such talk does not precede or preclude politics. For the babble emanating from this discursive space is a political process without end, and without a (unified collective) subject. In an overdetermined world conversations are interventions/actions/changes in and of themselves, no matter whether they do or do not also give rise to further planned interventions.

ACKNOWLEDGEMENTS

A longer version of this paper was published under the title '"Stuffed if I know!" Reflections on post-modern feminist social research', in *Gender, Place and Culture* 1(2) (1994): 205–24. During the writing of the paper Julie Graham was supported by a fellowship from the Faculty of Arts, Monash University. The fieldwork upon which the paper is based was funded by a grant to Katherine Gibson from the Australian Research Council.

NOTES

1 Butler (1990) argues, for example, that the category 'woman' is but a fiction of coherence that serves to buttress the heterosexual contract.

2 She writes,

> 'the world from women's point of view' and 'the feminist standpoint' attempt resolution of the postfoundational epistemology problem by deriving from within women's experience the grounding for women's accounts. But this resolution requires suspending recognition that women's 'experience' is both thoroughly constructed and interpreted without end. Within feminist standpoint theory as well as much other modernist feminist theory, then, consciousness raising operates as feminism's epistemologically positivist moment.
>
> (Brown 1991: 72).

The discussion of 'women's experience' is, from this perspective, the creation of a discourse which imposes a fixed identity, rather than the uncovering of an unmediated truth.

3 Barrett talks of the 'turn to culture' within feminism as the interest in post-structuralist theory has prompted a shift away from 'the social sciences' preoccupation with things' towards words and language. She claims that '(a)cademically, the social sciences have lost their purchase within feminism and the rising star lies with the arts, humanities and philosophy' (1992: 204–5).

4 The research findings will not be reported here. They have been written up in a number of research reports and articles (see Gibson 1991a, 1991b, 1992a, 1992b, 1993, and Gibson-Graham 1994a, 1994b).

5 Metcalfe (1987) argues that the exclusion of women from the coal-mining workforce in nineteenth-century Britain marked the active introduction (largely by the male union movement) of structured gender divisions within coal-mining communities. These divisions were transplanted to the Australian industry. In the terms of this discussion, this historical precedent provided the conditions under which miners' wives could only ever be constituted as Other to the working-class miner.

6 Four of the stories I collected are written up in Gibson-Graham (1994a).

7 I suspected that the traditional representation of 'mining town women' had a use within the discursive space occupied by social workers and service providers. The image of mining town women as individualistic non-joiners was a handy categorization which justified the interventionist activities of service providers and staff wives (often the same people in the smaller towns) who acted as gatekeepers for all social activity and any gender-based politics in the towns.

At the same time I was aware that the left representation of 'miner's wife' had a use within the discursive space occupied by union leaders, labour historians and socialist feminists. The image of a solid supporter of 'the men' and upholder of hard-won conditions valued and romanticized the contribution of women, and established the nobility of the miners' class struggle. For the wives of some miners this representation was a welcome reward for toeing the line in an important but self-effacing way. Women were accorded the accolades befitting true working-class warriors when they willingly subordinated their lives to the cause of jobs (for men), wages (for men), political rights (for men) or lower taxes (for men). But it seemed clear to me that these accolades would not be forthcoming if women overstepped some invisible mark and, for instance, interfered in wage negotiations or shiftwork changes when major disputes were not in the offing, or led a movement for paid work (for women), or wages for housework.

8 It was in the face of this possibility that Spivak's strategic essentialism was proposed (in an interview with Grosz 1984). Barrett cautions: 'Feminists recognize that the "naming" of women and men occurs within an opposition that one would want to challenge and transform. Yet political silencing can follow from rejecting these categories altogether' (1991: 166).

9 Brown (1991) has argued that in countering postmodern social fragmentation feminists need to orient their political conversations 'towards diversity and the common, toward world rather than self' and she encourages us to engage in a 'conversion of one's knowledge of the world from a situated (subject) position into a public idiom' (1991: 80–1). In a similar vein Weedon emphasizes the public realm arguing that 'in order for a discourse to have a social effect, [it] must at least be in circulation' (1987: 110–11).

10 The discourse of 'the field' is one that is undergoing an interesting deconstruction within anthropology (see, for example, D'Amico-Samuels 1991). Not surprisingly, within geography, it is feminist geographers who have taken up this particular challenge (Nast *et al.* 1994).

11 In framing the research project I was conscious of the disciplinary power that men, unions, the companies and the social service gatekeepers exercised over miners' wives in these communities. Negotiating permission from husbands, the unions, the service providers and the mining companies for these women to be involved was itself an interesting political exercise.

12 The questionnaire was designed to elicit information around the following topics: Who is providing all the unpaid labour which supports the physical and emotional needs of the shiftworker? How much of the increased productivity gained by continuous production is being fuelled by an intensification of household labour? How does the increased tiredness and lack of weekends affect relations between workers and partners, workers and children and partners and children – and workers and workers? How might a better understanding of women's experience of their partner's shiftwork patterns help men and women alike? What general feelings did women have about their town?

13 For me the first workshop marked a transition in my relationship to the women I was employing, which was initially dominated by the hierarchy and differential power of the academic/housewife-childrearer and employer/employee relations. By the end of the workshop, one of the women who was most into differentiating herself and her particular experience from that of the others (especially because she was quite happy with her life and felt that others' complaints didn't ring true for

her) expressed the view that even though I was a doctor I was really just one of them. This moment of identification referred primarily to one of my many subject positions, that of being a mother of small children, someone who could share in tales of childbirth, sleepless nights and the irrational frustrations of mothering. On the basis of this dimension of similarity I was somehow legitimated in her eyes, my power defused and her acceptance of me granted. I was homogenized and accepted into the fictional but collective unity 'mother'. In a different way Joanne (who was not a mother) was identified as a 'local' (that is, non-metropolitan), someone who experienced isolation, car breakdowns on outback roads and harsh climatic conditions, and partially accepted on that basis.

14 The Queensland branch of the United Mineworkers Federation has printed 5,000 copies of *Different Merry-Go-Rounds: Families, Communities and the 7-Day Roster* for distribution to its members and to the communities of Central Queensland.

CONCLUSION

Nancy Duncan

As I stated in the 'Introduction', the authors of this volume contribute in various ways to the feminist project of embodying, engendering and embedding knowledge claims and social research in the material context of space and place. This situating and specifying of theory and research is seen as necessary by feminists who have come to question concepts which pass as universal and disinterested, but turn out to refer to something much more particular and interested, usually privileged males. Allegedly gender-neutral concepts are too often based on an unstated masculine norm. A first step towards the solution to this problem is thought to consist in contextualizing and revealing the historical, cultural and gendered specificity of such universalist pretensions.

However, there are potential dangers in focusing on difference rather than identity, and specificity rather than generality. One of the more obvious problems concerns the defining of equality, which of course has long been a feminist goal. If one is to take difference and contingency seriously, equality must be defined in ways that do not assume homogeneity. As Anne Phillips has said: 'we cannot do without a notion of what human beings have in common; we can and must do without a unitary standard against which they are all judged' (1993: 66).

A material (and social) environment can be constructed which achieves greater equality of mobility and access by accounting for relative difference. What I mean by this is a levelling of the playing field, so to speak, by attempting to ensure that the social and physical environment itself does not unnecessarily handicap those who do not match a particular norm. An example would be a degree of equality in the workplace achieved through the granting of paid maternity and paternity leave which, while recognizing differences in family responsibilities, does so in the interest of equal treatment of all workers.

Judith Butler wishes to salvage some notion of universality by suggesting the concept itself be relieved of 'its foundational weight in order to render it a site of permanent political contest' (1992: 8). Here we can see that inclusion need not mean assimilation or co-option, but also that the recognition of difference need not lead to inequality. However, as we have seen, universalizing thought always courts the danger of falling into the trap of overgeneralized categories. What is important then, is to be ever vigilant in looking out for unintentional exclusions – for masculinism and cultural imperialism in our categories.

Butler goes on to say that a category such as woman should be a site of permanent contestation: 'Paradoxically, it may be that only through releasing the category of women from a fixed referent that something like "agency" becomes possible' (1992: 16). Contrary to much of the thinking around identity politics, a stable and unified identity is not a necessary basis for a progressive politics. Political agency can be effectively based on non-exclusionary, hetero-geneous categories. Accordingly one of the goals of this volume is to show that the categories of gender and sexuality do not map neatly onto one another and are sites of contestation and resistance against exclusions and dichotomizing tendencies. Exposing false universalist claims, however, need not lead us to turn our backs on Enlightenment ideals of social justice and universal human rights. These ideals must be contested whenever, and to the extent that, they can be exposed as implicitly ethnocentric, androcentric or exclusionary in any other way.

Geographers and others interested in place and historical specificity are well equipped to look at the importance of place to the construction of gender and sexuality difference and differences within these differences. However, their orientation toward place specificity has sometimes wrongly led them to conflate this with localism, based on a romantic, nostalgic or aesthetic sense of place. Here I refer to localism as a component of individualism which posits the rights of a theoretically free, equal, disembodied, gender-neutral, homogeneous group of individuals who come together at the local level to determine the interests of their specific community. Such localism and locally based politics can become myopic, turning attention away from regional or global political economic processes which structure inequalities. Thus, inappropriately localized or privatized solutions to problems may be sought. These often turn out to be either exclusionary or ineffective.

In her chapter Rose states that the idea of grounding or contextualizing feminist theory in so-called 'real' material space is a masculinist performance of power. Her alternative notions of spatiality indeed challenge not only conventional academic norms, but also might be considered a critique of many of the other chapters in this book. I agree that there is such a danger, especially if, as I have said, such grounding leads to localism and exclusion. Taking Rose's warning (in perhaps a more limited way than she intends) I suggest that material and discursive spatiality can be repoliticized and opened up to a more heterogeneous public. Rather than claiming space for a group, as in territorially based politics, moveable sites of resistance against exclusionary practices can break open such performances of power.

I am thinking here of something like Deleuze and Guattari's 'state' or 'striated' space (hold the fort) versus 'nomad' or 'smooth' space (hold the street) (see Massumi 1992: 6). In making a distinction between territories and smooth spaces I also draw inspiration from the title of Tim Davis's (1995) paper entitled 'Gay territories and queer spaces'. Here I interpret gay to mean a relatively stable identity based in part upon sexuality which can be mapped onto stable and relatively fixed locations, and queer to refer to a destabilizing oppositional politics of sexuality which is associated with a fluid spatiality and multiplying and moveable sites of resistance.[1]

The general question which this book addresses is exactly how to bring about the feminist goal of a structural transformation which goes beyond simply making amendments to previously existing theory. The essays in this volume make some valuable suggestions which are primarily geographical in orientation. This reveals not only the particular interests and expertise of the authors, but reflects a much more widely recognized need within feminism today to specify by situating the often too general, allegedly neutral claims of social and cultural theory. Thus the authors of the essays in this book have offered some of their ideas about how to engender and contextualize knowledge claims through a repoliticized geographical imagination.

NOTE

1 An earlier version of Davis's paper was delivered at Syracuse University in the symposium entitled 'Place, Space and Gender' from which most of the papers in this volume were drawn. A printed version of the paper under a different title appears in *Mapping Desire* edited by David Bell and Gill Valentine (Davis, 1995: 284–302).

REFERENCES

INTRODUCTION

Alcoff, Linda and Potter, Elizabeth (eds) (1993) *Feminist Epistemologies*, New York: Routledge.

Anderson, Kay (1991) *Vancouver's Chinatown: Racial Discourse in Canada, 1875–1980*, Montreal: McGill-Queen's University Press.

Bhabha, Homi K. (1994) *The Location of Culture*, London: Routledge.

Bordo, Susan (1987) *The Flight to Objectivity: Essays on Cartesianism and Culture*, New York: SUNY Press.

Brown, Wendy (1995) *States of Injury: Power and Freedom in Late Modernity*, Princeton, NJ: Princeton University Press.

Butler, Judith (1990) *Gender Trouble: Feminism and the Subversion of Identity*, New York: Routledge.

—— (1992) 'Contingent foundations: feminism and the question of "postmodernism"', in Judith Butler and Joan Scott (eds) *Feminists Theorize the Political*, New York: Routledge.

Crary, Jonathan (1993) *Techniques of the Observer*, Cambridge, MA: MIT Press.

England, Kim (1994) 'Getting personal: reflexivity, positionality and feminist research', *The Professional Geographer* 46(1): 80–9.

Gross, Elizabeth (1987) 'Conclusion: what is feminist theory?' in Carole Pateman and Elizabeth Gross (eds) *Feminist Challenges: Social and Political Theory*, pp. 190–204, Boston: Northeastern University Press.

Haraway, Donna (1991) 'Situated knowledges: the science question in feminism and the privilege of partial perspective', in *Simians, Cyborgs, and Women*, New York: Routledge.

Harding, Sandra (1986) *The Science Question in Feminism*, Ithaca, NY: Cornell University Press.

—— (1991) *Whose Science? Whose Knowledge?* Ithaca, NY: Cornell University Press.

—— (1993) 'Rethinking standpoint theory: "What is strong objectivity?"' in Linda Alcoff and Elizabeth Potter (eds) *Feminist Epistemologies*, pp. 49–82, New York: Routledge.

Hartsock, Nancy (1983) 'The feminist standpoint: developing the ground for a specifically feminist historical materialism', in Sandra Harding and Merrill Hintikka (eds) *Discovering Reality: Feminist Perspectives on Epistemology*, Dordrecht: Reidel.

Harvey, David (1993) 'Class relations, social justice and the politics of difference', in Judith Squires (ed.) *Principled Positions: Postmodernism and the Rediscovery of Value*, London: Lawrence & Wishart.

Lloyd, Genevieve (1984) *The Man of Reason: 'Male' and 'Female' in Western Philosophy*, Minneapolis: University of Minnesota Press.

Merchant, Carolyn (1983) *The Death of Nature: Women, Ecology and the Scientific Revolution*, San Francisco: Harper.

Spivak, Gayatri Chakravorty (1988) 'Can the subaltern speak?' in C. Nelson and L.

Grossberg (eds) *Marxism and the Interpretation of Culture*, pp. 271–313, Basingstoke: Macmillan Education.

Valentine, Gill (1993) '(Hetero)sexing space: lesbian perceptions and experiences of everyday spaces', *Environment and Planning D: Society and Space* 11: 395–413.

Young, Iris Marion (1987) 'Impartiality and the civic public: some implications of feminist critiques of moral and political theory', in Seyla Benhabib and Drucilla Cornell (eds) *Feminism as Critique*, pp. 56–76, Minneapolis: University of Minnesota Press.

CHAPTER 1
FEMINIST THEORY AND SOCIAL SCIENCE

Alcoff, Linda Martín (1996) 'Is the feminist critique of reason rational?' *Philosophical Topics*: forthcoming.

Barrett, Nancy S. (1981) 'How the study of women has restructured the discipline of economics', in Elizabeth Langland and Walter Grove (eds) *A Feminist Perspective in the Academy*, pp. 101–9, Chicago: University of Chicago Press.

Bell, Linda (1983) *Visions of Women*, Clifton, NJ: Humana Press.

Braidotti, Rosi (1991) *Patterns of Dissonance*, New York: Routledge.

Flax, Jane (1990) *Thinking Fragments: Psychoanalysis, Feminism and Postmodernism in the Contemporary West*, Los Angeles: University of California Press.

Gross, Elizabeth (1987) 'What is feminist theory?' in Carole Pateman and Elizabeth Gross (eds) *Feminist Challenges: Social and Political Theory*, pp. 190–204, Boston: Northeastern University Press.

Grosz, Elizabeth (1993) 'Bodies and knowledges: feminism and the crisis of reason', in L. Alcoff and E. Potter (eds) *Feminist Epistemologies*, pp. 187–216, New York: Routledge.

Harding, Sandra (1991) *Whose Science? Whose Knowledge?* Ithaca, NY: Cornell University Press.

Irigaray, Luce (1993) *An Ethics of Sexual Difference*, transl. Carolyn Burke and Gillian C. Gill, Ithaca, NY: Cornell University Press.

Keohane, Nannerl O. (1981) 'Speaking from silence: women and the science of politics', in Elizabeth Langland and Walter Grove (eds) *A Feminist Perspective in the Academy*, pp. 86–100, Chicago: University of Chicago Press.

Kittay, Eva Feder (1987) *Metaphor: Its Cognitive Force and Linguistic Structure*, Oxford: Clarendon Press.

—— (1988) 'Woman as metaphor', *Hypatia* 3(2): 63–86.

Le Doeuff, Michelle (1989) *The Philosophical Imaginary*, London: Athlone Press.

Lloyd, Genevieve (1984) *The Man of Reason: 'Male' and 'Female' in Western Philosophy*, Minneapolis: University of Minnesota Press.

Longino, Helen (1990) *Science as Social Knowledge: Values and Objectivity in Scientific Inquiry*, Princeton, NJ: Princeton University Press.

Lovibond, Sabina (1994) 'Feminism and the "crisis of rationality"', *New Left Review* 207 (Sept./Oct.): 72–86.

Ong, Aihwa (1988) 'Colonialism and modernity: feminist re-presentations of women in non-Western societies', *Inscriptions* 3(4): 79–93.

Plato (1961) *Collected Dialogues*, edited by Edith Hamilton and Huntington Cairns, Princeton, NJ: Princeton University Press.

Rubin, Gayle (1975) 'The traffic in women: notes on the "political economy" of sex', in Rayna Reiter (ed.) *Toward an Anthropology of Women*, pp. 157–210, New York: Monthly Review Press.

CHAPTER 2 SPATIALIZING FEMINISM

Anderson, B. (1983) *Imagined communities: Reflections on the origin and spread of nationalism*, London: Verso.

Bhabha, H. (1994) *The Location of Culture*, London: Routledge.

Brydon, L. and Chant, S. (1989) *Women in the Third World: Gender Issues in Rural and Urban Areas*, London: Edward Elgar.

Butler, J. (1990) *Gender Trouble: Feminism and the Subversion of Identity*, London: Routledge.

Carter, E., Donald, J. and Squires, J. (eds) (1993) *Space and Place: Theories of Identity and Location*, London: Lawrence & Wishart.

Castells, M. (1989) *The Informational City*, Oxford: Blackwell.

Foucault, M. (1986) 'Of other spaces', *diacritics* Spring

Geertz, C. (1988) *Works and Lives: The Anthropologist as Author*, Cambridge: Polity.

Giddens, A. (1990) *The Consequences of Modernity*, Cambridge: Polity.

Gilroy, P. (1993) *The Black Atlantic: Modernity and Double Consciousness*, London: Verso.

Hall, S. (1991) 'The local and the global: globalization and ethnicity', in A. King (ed.) *Culture, Globalization and the World System*, pp. 19–39, London: Macmillan.

—— (1994) 'The question of cultural identity', in *The Polity Reader in Cultural Theory*, pp. 119–25, Cambridge: Polity.

Hanson, S. and Pratt, G. (1995) *Gender, Place and Work*, London: Routledge.

Haraway, D. (1991) *Simians, Cyborgs and Women: The Reinvention of Nature*, London: Free Association Books.

Hartsock, N. (1990) 'Foucault on power: a theory for women', in L.J. Nicholson (ed.) *Feminism/Postmodernism*, London: Routledge.

Harvey, D. (1989) *The Condition of Postmodernity*, Oxford: Blackwell.

Jameson, F. (1991) *Postmodernism or, The Cultural Logic of Late Capitalism*, Durham, NC: Duke University Press.

Katz, C. and Monk, J. (eds) (1994) *Full Circles: Geographies of Women Over the Life Course*, London: Routledge.

Laclau, E. (1990) *New Reflections on the Revolution of Our Time*, London: Verso.

Leidner, R. (1993) *Fast Food, Fast Talk*, Los Angeles and San Francisco: University of Berkeley Press.

Little, J., Peake, L. and Richardson, P. (eds) (1988) *Women in Cities*, London: Macmillan.

Mascia-Lees, F., Sharpe, P. and Cohen, C.B. (1989) 'The postmodernist turn in anthropology: cautions from a feminist perspective', *Signs* 15: 7–33.

Massey, D. (1992) 'Politics and space/time', *New Left Review* 196: 65–84.

—— (1993) 'Power-geometry and a progressive sense of place', in J. Bird, B. Curtis, G. Robertson and L. Tickner (eds) *Mapping the Futures: Local Culture, Global Change*, London: Routledge.

Massey, D. and Allen, J. (eds) (1984) *Geography Matters!* Cambridge: Cambridge University Press.

McDowell, L. (1993) 'Space, place and gender relations: Part II. Identity, difference, feminist geometries and geographies', *Progress in Human Geography* 17: 305–18.

Minh-ha, T.T. (1987) *Woman, Native, Other*, Bloomington, IN: Indiana University Press.

Mohanty, C.T. (1991) 'Cartographies of struggle: Third World women and the politics of feminism', in C.T. Mohanty, A. Russo and L. Torres (eds) *Third World Women and the Politics of Feminism*, Bloomington, IN: Indiana University Press.

Momsen, J. and Kinnaird, V. (eds) (1994) *Different Places, Different Voices: Gender and Development in Africa, Asia and Latin America*, London: Routledge.

Momsen, J. and Townsend, J. (eds) (1987) *Geography and Gender in the Third World*, London: Hutchinson.

Moore, H.L. (1994) *A Passion for Difference*, Cambridge: Polity.

Pringle, R. (1989) *Secretaries Talk*, London: Verso.

Probyn, E. (1994) *Sexing the Self*, London: Routledge.

Smith, N. (1993) 'Homeless/global: scaling places', in J. Bird, B. Curtis, G. Robertson and L. Tickner (eds) *Mapping the Futures: Local Culture, Global Change*, London: Routledge.

Soja, E. (1989) *Postmodern Geographies: The Reassertion of Space in Social Theory*, London: Verso.

Squires, J. (ed.) (1993) *Principled Positions: Postmodernism and the Rediscovery of Value*, London: Lawrence & Wishart.

Tivers, J. (1985) *Women Attached: The Daily Lives of Women with Children*, London: Croom Helm.

Young, I.M. (1990) 'The ideal of community and the politics of difference', in L.J. Nicholson (ed.) *Feminism/Postmodernism*, London: Routledge.

CHAPTER 3 RE: MAPPING SUBJECTIVITY

Benjamin, Jessica (1988) *The Bonds of Love: Psychoanalysis, Feminism and the Problem of Domination*, New York: Pantheon.

Champlain, Samuel de (1907) *Voyages of Samuel de Champlain 1604–1618*, edited by W.L. Grant, Original Narratives of Early American History, New York: Scribner.

Fox Keller, Evelyn (1985) *Reflections on Gender and Science*, New Haven, CT: Yale University Press.

Freud, Sigmund (1923/1961) *The Ego and the Id*, in *Standard Edition* 19, pp. 3–66, transl. and edited by James Strachey, London: Hogarth.

—— (1930/1961) *Civilization and Its Discontents*, transl. and edited by James Strachey, New York: Norton.

Harding, Sandra (1986) *The Science Question in Feminism*, Ithaca, NY: Cornell University Press.

Harley, J.B. (1988) 'Silences and secrecy: the hidden agenda of cartography in Early Modern Europe', *Imago Mundi* 40: 57–76.

—— (1989) 'Deconstructing the map', *Cartographica* 26(2): 1–20.

—— (n.d.) 'Victims of a map: New England cartography and the Native Americans', unpublished manuscript.

Helgerson, Richard (1986) 'The land speaks: cartography, chorography, and subversion in Renaissance England', *Representations* 16: 51–85.

Jameson, Fredric (1984) 'Postmodernism, or the cultural logic of late capitalism', *New Left Review* 146: 53–92.

—— (1988) 'Cognitive mapping', in Cary Nelson and Lawrence Grossberg (eds) *Marxism and the Interpretation of Culture*, pp. 347–57, Chicago: University of Illinois Press.

—— (1991) *Postmodernism or, The Cultural Logic of Late Capitalism*, Durham, NC: Duke University Press.

JanMohamed, Abdul (1986) 'The economy of Manichean allegory', in Henry Louis Gates Jr (ed.) *'Race', Writing and Difference*, Chicago: Chicago University Press.

Lacan, J. (1977) 'Of the subject of certainty', in Jacques-Alain Miller (ed.) *Four Fundamental Concepts of Psychoanalysis*, transl. Alan Sheridan, New York: Norton.

Mazey, Mary Ellen and Lee, David R. (1983) *Her Space, Her Place: A Geography of Women*, Resource Publications in Geography, Washington, DC: Association of American Geographers.

Mitchard, Jacqueline (1989) 'Men still fail to pick up on all the pieces', *Milwaukee Journal* 30 July: G1.

Rubinstein, C. (1980) 'Survey report: how Americans view vacations', *Psychology Today* 14: 62–76.

Spivak, Gayatri (1991) 'Three women's texts and a critique of imperialism', in Robyn R. Warhol and Diane Price Herndl (eds) *Feminisms: An Anthology of Literary Theory and Criticism*, pp. 798–814, New Brunswick: Rutgers University Press.

Tuan, Y.F. (1974) *Topophilia: A Study of Environmental Perception, Attitudes, and Values*, Englewood Cliffs, NJ: Prentice Hall.

Turner, Frederick (1980) *Beyond Geography: The Western Spirit Against the Wilderness*, New York: Viking.

Vaca, Cabeza de (1961) *Adventures in the Unknown Interior of America*, transl. and edited by Cyclone Covey, Albuquerque: University of New Mexico Press.

CHAPTER 4 AS IF THE MIRRORS HAD BLED

Agnew, J. (1993) 'Representing space: space, scale and culture in social science', in J.S. Duncan and D. Ley (eds) *Place/Culture/Representation*, pp. 251–71, London: Routledge.

Bhabha, H. (1990) 'The other question: difference, discrimination and the discourse of colonialism', in R. Ferguson *et al.* (eds) *Out There: Marginalization and Contemporary Cultures*, pp. 71–87, New York and Cambridge, MA: New Museum of Contemporary Art and MIT Press.

Bondi, L. (1992) 'Gender symbols and urban landscapes', *Progress in Human Geography* 16(2): 157–70.

—— (1993) 'Locating identity politics', in M. Keith and S. Pile (eds) *Place and the Politics of Identity*, pp. 84–103, London: Routledge.

Braidotti, R. (1991) *Patterns of Dissonance: A Study of Women in Contemporary Philosophy*, Cambridge: Polity.

Brennan, T. (1991) 'An impasse in psychoanalysis and feminism', in S. Gunew (ed.) *A Reader in Feminist Knowledge*, pp. 114–38, London: Routledge.

Butler, J. (1990) *Gender Trouble: Feminism and the Subversion of Identity*, London: Routledge.

—— (1993) *Bodies That Matter: On the Discursive Limits of 'Sex'*, London: Routledge.

Christopherson, S. (1989) 'On being outside "the project"', *Antipode* 21(1): 83–9.

Daniels, S. and Cosgrove, D. (1993) 'Spectacle and text: landscape metaphors in cultural geography' in J.S. Duncan and D. Ley (eds) *Place/Culture/Representation*, pp. 57–77, London: Routledge.

Derrida, J. (1979) *Writing and Difference*, transl. A. Bass, London: Routledge & Kegan Paul.

Elliott, A. (1992) *Social Theory and Psychoanalysis in Transition*, Oxford: Blackwell.

Fuss, D. (1989) '"Essentially speaking": Luce Irigaray's language of essence', *Hypatia* 3(3): 62–80.

Gallop, J. (1988) *Thinking Through the Body*, New York: Columbia University Press.

Gregory, D. (1994) *Geographical Imaginations*, Oxford: Blackwell.

Grosz, E. (1989) *Sexual Subversions: Three French Feminists*, Sydney: Allen & Unwin.

Hanson, S. (1992) 'Presidential Address – Geography and feminism: worlds in collision?' *Annals of the Association of American Geographers* 82(4): 569–86.

Haraway, D. (1991) *Simians, Cyborgs and Women: The Reinvention of Nature*, London: Free Association Books.

Harvey, D. (1989) *The Condition of Postmodernity*, Oxford: Blackwell.

—— (1993) 'From space to place and back again: reflections on the condition of post-modernity', in J. Bird, B. Curtis, G. Robertson and L. Tickner (eds) *Mapping the Futures: Local Cultures, Global Change*, pp. 3–29, London: Routledge.

Holmlund, C. (1989) 'I love Luce: the lesbian, mimesis and masquerade', *New Formations* 9: 105–23.

Irigaray, L. (1985a) *This Sex Which Is Not One*, transl. C. Porter, Ithaca, NY; Cornell University Press.

—— (1985b) *Speculum of the Other Woman*, transl. G.C. Gill, Ithaca, NY: Cornell University Press.

—— (1991) *The Irigaray Reader*, edited by M. Whitford, Oxford: Blackwell.

—— (1992) *Elemental Passions*, transl. J. Collie and J. Still, London: Athlone Press.

—— (1993a) *An Ethics of Sexual Difference*, transl. C. Burke and G.C. Gill, London: Athlone Press.

—— (1993b) *Sexes and Genealogies*, transl. G.C. Gill, New York: Columbia University Press.

—— (1993c) *Je, Tu, Nous: Toward a Culture of Difference*, transl. A. Martin, London: Routledge.

Keith, M. and Pile, S. (1993) 'Introduction: the politics of place', in M. Keith and S. Pile (eds) *Place and the Politics of Identity*, pp. 1–21, London: Routledge.

Lagopoulos, A.P. (1993) 'Postmodernism, geography, and the social semiotics of space', *Environment and Planning D: Society and Space* 11(3): 255–78.

Massey, D. (1993) 'Politics and space/time', in M. Keith and S. Pile (eds) *Place and the Politics of Identity*, pp. 141–61, London: Routledge.

Merrifield, A. (1993) 'Place and space: a Lefebvrian reconciliation', *Transactions of the Institute of British Geographers* 18(4): 516–31.

Morris, M. (1988) *The Pirate's Fiancée: Feminism, Reading, Postmodernism*, London: Verso.

—— (1992) 'The man in the mirror', *Theory, Culture & Society* 10(2–3): 253–79.

Pratt, G. (1992) 'Spatial metaphors and speaking positions', *Environment and Planning D: Society and Space* 10(3): 241–3.

Ragland-Sullivan, E. (1992) 'The imaginary', in E. Wright (ed.) *Feminism and Psychoanalysis: A Critical Dictionary*, pp. 173–6, Oxford: Blackwell.

Reichert, D. (1992) 'On boundaries', *Environment and Planning D: Society and Space* 10(1): 87–98.

—— (1994) 'Woman as utopia', *Gender, Place and Culture* 1(1): 91–102.

Rose, G. (1993) *Feminism and Geography: The Limits of Geographical Knowledge*, Cambridge: Polity.

Smith, N. (1984) *Uneven Development: Nature, Capital and the Production of Space*, Oxford: Blackwell.

—— (1993) 'Homeless/global: scaling places', in J. Bird, B. Curtis, G. Robertson and L. Tickner (eds) *Mapping the Futures: Local Culture, Global Change*, pp. 87–119, London: Routledge.

Smith, N. and Katz, C. (1993) 'Grounding metaphor: towards a spatialized politics', in M. Keith and S. Pile (eds) *Place and the Politics of Identity*, pp. 67–83, London: Routledge.

Soja, E. (1987) 'The postmodernization of geography: a review', *Annals of the Association of American Geographers* 77(2): 289–94.

Soja, E. and Hooper, B. (1993) 'The spaces that difference makes: some notes on the geographical margins of the new cultural politics', in M. Keith and S. Pile (eds) *Place and the Politics of Identity*, pp. 183–205, London: Routledge.

Whitford, M. (1986) 'Luce Irigaray and the female imaginary: speaking as a woman', *Radical Philosophy* 43: 3–8

—— (1991) *Luce Irigaray: Philosophy in the Feminine*, London: Routledge.

CHAPTER 5 RE-CORPOREALIZING VISION

Bordo, Susan (1987) *The Flight to Objectivity: Essays on Cartesianism and Culture*, New York: SUNY Press.

Braverman, Harry (1974) *Labor and Monopoly Capital*, New York: Monthly Review Press.

Collins, Patricia Hill (1991a) *Black Feminist Thought*, New York: Routledge.

Cosgrove, Denis (1985) 'Prospect, perspective and the evolution of the landscape idea', *Transactions of the Institute of British Geographers* 10: 45–62.

Crary, Jonathan (1988) 'Modernizing vision', in Hal Foster (ed.) *Vision and Visuality*, Seattle: Bay Press.

—— (1993) *The Techniques of the Observer*, Cambridge, MA: MIT Press.

Grosz, Elizabeth (1994) *Volatile Bodies: Toward a Corporeal Feminism*, Bloomington: Indiana University Press.

Haraway, Donna (1991a) 'Animal sociology and a natural economy of the body politic', in *Simians, Cyborgs and Women: The Reinvention of Nature*, New York: Routledge.

—— (1991b) 'Situated knowledges: the science question in feminism and the privilege of partial perspective', in *Simians, Cyborgs and Women: The Reinvention of Nature*, New York: Routledge.

Harley, Brian (1989) 'Deconstructing the map', *Cartographica* 26(2): 1–20.

hooks, bell (1990) *Yearning: race, gender and cultural politics*, Boston: South End Press.

Hoover, Herbert and Hoover, Lou (1950) *Georgius Agricola: De Re Metallica*, New York: Dover Publications Inc.

Jacobs, Harriet (1988) *Incidents in the Life of a Slave Girl*, Oxford: Oxford University Press.

Lefebvre, Henri (1991) *The Production of Space*, Oxford: Blackwell.

Merchant, Carolyn (1983) *The Death of Nature: Women, Ecology and the Scientific Revolution*, San Francisco: Harper.

Mintz, Sidney W. (1985) *Sweetness and Power: The Place of Sugar in Modern History*, New York: Penguin Books.

Mitchell, Timothy (1988) *Colonising Egypt*, Cambridge: Cambridge University Press.

Natter, W. and Jones, J.P. (1993) 'Pets or meat: class, ideology, and space in *Roger and Me*', *Antipode* 25: 140–58.

Rose, Gillian (1993a) 'The geographical imagination', in *Feminism and Geography*, pp. 62–86, Minneapolis: University of Minnesota Press.

—— (1993b) 'A politics of paradoxical space', in *Feminism and Geography*, pp. 137–61, Minneapolis: University of Minnesota Press.

Schiebinger, Londa (1989) *The Mind Has No Sex? Women in the Origins of Modern Science*, Cambridge, MA: Harvard University Press.

Wigley, Mark (1992) 'Untitled: the housing of gender', in *Sexuality and Space*, pp. 327–89, New York: Princeton Architectural Press.

CHAPTER 6 GENDERING NATIONHOOD

Anderson, Benedict (1991) *Imagined Communities: Reflections on the Origin and Spread of Nationalism*, 2nd edn (original, 1983). New York: Verso.

Antić, Milica (1991) 'Democracy between tyranny and liberty: women in post-"socialist" Slovenia', *Feminist Review* 39: 149–54.

Bennington, Geoffrey (1990) 'Postal politics and the institution of the nation', in Homi K. Bhabha (ed.) *Nation and Narration*, pp. 121–37, New York: Routledge.

Butler, Judith (1990) *Gender Trouble: Feminism and the Subversion of Identity*, New York: Routledge.

Campbell, David (1992) *Writing Security: United States Foreign Policy and the Politics of Identity*, Minneapolis: University of Minnesota Press.

Chatterjee, Partha (1993) *The Nation and Its Fragments*. Princeton, NJ: Princeton University Press.

Dalby, Simon (1994) 'Gender and geopolitics: reading security discourse in the new world order', *Environment and Planning D: Society and Space* 12(5): 525–42.

Dölling, Irene (1991) 'Between hope and hopelessness: women in the GDR after the "Turning Point"', *Feminist Review* 39: 3–15.

Drakulić, Slavenka (1993) 'Women and the new democracy in the former Yugoslavia', in Nanette Funk and Magda Mueller (eds) *Gender Politics and Post-Communism: Reflections from Eastern Europe and the Former Soviet Union*, pp. 123–30, New York: Routledge.

Einhorn, Barbara (1991) 'Where have all the women gone? Women and the women's movement in East Central Europe', *Feminist Review* 39: 16–36.

—— (1993) *Cinderella Goes to Market: Citizenship, Gender and Women's Movements in East Central Europe*, New York: Verso.

Enloe, Cynthia (1989) *Bananas, Beaches and Bases: Making Feminist Sense of International Relations*, Berkeley: University of California Press.

—— (1993) *The Morning After: Sexual Politics at the End of the Cold War*, Berkeley: University of California Press.

Fábián, Katalin (1993) 'The political aspects of women's changing status in Central and Eastern Europe', in *Proceedings of the 1993 Maxwell Colloquium*, pp. 73–80, Maxwell School of Citizenship and Public Affairs, Syracuse University.

Foucault, Michel (1980) 'The history of sexuality', in *Power/Knowledge: Selected*

Interviews and Other Writings 1972–1977 by Michel Foucault, pp. 183–93, edited by Colin Gordon, London: Harvester Wheatsheaf.

Funk, Nanette (1993) 'Feminism East and West', in Nanette Funk and Magda Mueller (eds) *Gender Politics and Post-Communism: Reflections from Eastern Europe and the Former Soviet Union*, pp. 318–30, New York: Routledge.

Funk, Nanette and Mueller, Magda (eds) (1993) *Gender Politics and Post-Communism: Reflections from Eastern Europe and the Former Soviet Union*, New York: Routledge.

Gorbachev, Mikhail (1987) *Perestroika: New Thinking for Our Country and the World*, London: Fontana.

Goven, Joanna (1993) 'Gender politics in Hungary: autonomy and antifeminism', in Nanette Funk and Magda Mueller (eds) *Gender Politics and Post-Communism: Reflections from Eastern Europe and the Former Soviet Union*, pp. 224–40, New York: Routledge.

Hauser, Ewa, Heynes, Barbara and Mansbridge, Jane (1993) 'Feminism in the interstices of politics and culture: Poland in transition', in Nanette Funk and Magda Mueller (eds) *Gender and Post-Communism: Reflections from Eastern Europe and the Former Soviet Union*, pp. 257–73, New York: Routledge.

Jankowska, Hanna (1991) 'Abortion, Church and politics in Poland', *Feminist Review* 39: 174–81.

Johnson, Nuala (1995) 'The renaissance of nationalism', in R.J. Johnston, P.J. Taylor and W.J. Watts (eds) *Geographies of Global Change: Remapping the World in the Late Twentieth Century*, Oxford: Blackwell.

Kiss, Yudit (1991) 'The second "No": women in Hungary', *Feminist Review* 39: 49–57.

Marston, Sally (1990) 'Who are "the people"? Gender, citizenship, and the making of the American nation', *Environment and Planning D: Society and Space* 8: 449–58.

McClintock, Anne (1993) 'Family feuds: gender, nationalism and the family', *Feminist Review* 44: 61–80.

Milić, Andjelka (1993) 'Women and nationalism in the former Yugoslavia', in Nanette Funk and Magda Mueller (eds) *Gender and Post-Communism: Reflections from Eastern Europe and the Former Soviet Union*, pp. 109–22, New York: Routledge.

Molyneux, Maxine (1991) Interview with Anastasya Posadskaya (25 September 1990). *Feminist Review* 39: 133–40.

Mouffe, Chantal (1992) 'Feminism, citizenship and radical democratic politics', in Judith Butler and Joan Scott (eds) *Feminists Theorize the Political*, pp. 369–84, New York: Routledge.

Nash, Catherine (1994) 'Remapping the body/land: new cartographies of identity, gender, and landscape in Ireland', in Alison Blunt and Gillian Rose (eds) *Writing Women and Space: Colonial and Postcolonial Geographies*, pp. 227–50, New York: Guilford Press.

Radhakrishnan, R. (1992) 'Nationalism, gender, and the narrative of identity', in Andrew Parker, Mary Russo, Doris Sommer and Patricia Yaeger (eds) *Nationalisms and Sexualities*, pp. 77–95, New York: Routledge.

Renan, Ernest (1990) 'What is a nation?' in Homi K. Bhabha (ed.) *Nation and Narration*, pp. 8–22, New York: Routledge.

Sparke, Matthew (1994) 'Writing on patriarchal missiles: the chauvinism of the "Gulf War" and the limits of critique', *Environment and Planning A* 26: 1061–89.

Spivak, Gayatri Chakravorty and Guha, Ranajit (eds) (1988) *Selected Subaltern Studies*, New York: Oxford University Press.

Tickner, J. Ann (1992) *Gender and International Relations*, New York: Columbia University Press.

Todova, Maria (1993) 'The Bulgarian case: women's issues or feminist issues?' in Nanette Funk and Magda Mueller (eds) *Gender Politics and Post-Communism: Reflections from Eastern Europe and the Former Soviet Union*, pp. 30–8, New York: Routledge.

Watson, Peggy (1993) 'Eastern Europe's silent revolution: gender', *Sociology* 27(3): 471–87.

Young, Iris M. (1990) 'The ideal of community and the politics of difference', in Linda J. Nicholson (ed.) *Feminism/Postmodernism*, pp. 300–23, New York: Routledge.

Yuval-Davis, Nira (1991) 'The citizenship debate: women, ethnic processes and the state', *Feminist Review* 39: 58–68.

CHAPTER 7 MASCULINITY, DUALISMS AND HIGH TECHNOLOGY

Bourdieu, P (1977) *Outline of a Theory of Practice*, Cambridge Studies in Social Anthropology, Cambridge: Cambridge University Press.

Cockburn, C. (1985) *Machinery of Dominance: Women, Men and Technical Know-how*, London: Pluto Press.

de Beauvoir, S. (1949/1972) *The Second Sex*, transl. H.M. Parshley, Harmondsworth: Penguin.

Dinnerstein, D. (1987) *The Rocking of the Cradle and the Ruling of the World*, London: The Women's Press.

Easlea, B. (1981) *Science and Sexual Oppression: Patriarchy's Confrontation with Woman and Nature*, London: Weidenfield & Nicolson.

Halford, S. and Savage, M. (1995) 'Restructuring organisations, changing people: gender and restructuring in banking and local government', *Work, Employment and Society* 9(1): 97–122.

Hall, P. (1985) 'The geography of the fifth Kondratieff', in P. Hall and A. Markusen (eds) *Silicon Landscapes*, pp. 1–19, London, Allen & Unwin.

Hartsock, N. (1985) *Money, Sex and Power*, Boston: Northeastern University Press.

Henry, N and Massey, D. (1995), 'Competitive time-space in high technology', *Geoforum* 26(1): 49–64.

Ho, M.-W. (1993) *The Rainbow and the Worm: The Physics of Organisms*, London: World Scientific.

Keller, E.F. (1982) 'Feminism and science', *Signs: Journal of Women in Culture and Society* 7(3): 589–602.

—— (1985) *Reflections on Gender and Science*, New Haven CT: Yale University Press.

Kidder, T. (1982) *The Soul of a New Machine*, Harmondsworth: Penguin.

Lefebvre, H. (1991) *The Production of Space*, Oxford: Blackwell.

Lloyd, G. (1984) *The Man of Reason: 'Male' and 'Female' in Western Philosophy*, London: Methuen.

Massey, D. and Henry, N. (forthcoming) 'Time–spaces in and around high-tech'.

Massey, D., Quintas, P. and Wield, D. (1992) *High-tech Fantasies: Science Parks in Society, Science and Space*, London: Routledge.

Moore, H. (1986) *Space, Text and Gender*, Cambridge: Cambridge University Press.

Noble, D. (1992) *A World without Women: The Christian Clerical Culture of Western Science*, New York: Alfred A. Knopf.

O'Brien, M. (1981) *The Politics of Reproduction*, London: Routledge & Kegan Paul.

Sartre, J.-P. (1943) *Being and Nothingness*, transl. H.E. Barnes, London: Methuen, 1958.

Segal Quince & Partners (1985) *The Cambridge Phenomenon*, Cambridge: Segal Quince & Partners.

Turkle, S. (1984) *The Second Self: Computers and the Human Spirit*, London: Granada.

Wajcman, J. (1991) *Feminism Confronts Technology*, Cambridge: Polity Press.

CHAPTER 8 RENEGOTIATING GENDER AND SEXUALITY IN PUBLIC AND PRIVATE SPACES

Anzaldua, G. (1987) *The Borderlands/La Frontera New Mestiza*, San Francisco: Aunt Lute Books.

Bell, D. (1995) 'Perverse dynamics, sexual citizenship, and the transformation of intimacy', in D. Bell and G. Valentine (eds) *Mapping Desire*, London: Routledge.

Bell, D. and Valentine, G. (1995) 'Introduction' in D. Bell and G. Valentine (eds) *Mapping Desire*, London: Routledge.

Bell, D., Binnie, J., Cream, J. and Valentine, G. (1994) 'All hyped up and no place to go', *Gender, Place and Culture* 1(1): 31–47.

Bell, S. (1994) *Reading, Writing and Rewriting the Prostitute Body*, Bloomington, IL.: Indiana University Press.

Benhabib, S. (1992) *Situating the Self: Gender, Community and Postmodernism in Contemporary Ethics*, New York: Routledge.

Brownmiller, S. (1975) *Against Our Will: Men, Women and Rape*, London: Simon & Schuster.

Buel, S. (1988) 'Mandatory arrest for domestic violence', *Harvard Women's Law Reporter* 11.

Butler, J. (1990) *Gender Trouble: Feminism and the Subversion of Identity*, London: Routledge.

—— (1994) 'Gender as performance', *Radical Philosophy* 67: 32–9.

Calhoun, C. (ed.) (1993) *Habermas and the Public Sphere*, Cambridge, MA: MIT Press.

Cobbe, S. (1868) *Criminals, Idiots, Women, and Minors*, in J. Cohen and A. Arato (1992) *Civil Society and Political Theory*, Cambridge, MA: MIT Press.

Cohen, J. and Arato, A. (1992) *Civil Society and Political Theory*, Cambridge, MA: MIT Press.

Crenshaw, K. (1994) 'Mapping the margins: intersectionality, identity politics, and violence against women of color', in M. Fineman and R. Mykitiuk (eds) *The Public Nature of Private Violence*, New York: Routledge.

Cresswell, T. (1996) *In Place/Out of Place: Geography, Ideology and Transgression*, Minneapolis: University of Minnesota Press.

de Beauvoir, S. (1974) *The Second Sex*, Harmondsworth: Penguin. (First published 1949.)

Deleuze, G. and Guattari, F. (1987) *A Thousand Plateaus: Capitalism and Schizophrenia*, transl. B. Massumi, Minneapolis: University of Minnesota Press.

Edwards, S. (ed.) (1989) 'The extent of the problem: how widespread is domestic violence?' In *Policing Domestic Violence: Women, the Law and the State*, London: Sage.

Foucault, M. (1980) *Power/Knowledge: Selected Interviews and Other Writings, 1972–1977* edited by Colin Gordon, New York: Random House.

Fraser, N. (1992) 'Rethinking the public sphere: a contribution to the critique of actually existing democracy', in C. Calhoun (ed.) *Habermas and the Public Sphere*, Cambridge, MA: MIT Press.

Habermas, J. (1991) *The Structural Transformation of the Public Sphere*, Cambridge, MA: MIT Press.

Hammerton, A. J. (1992) *Cruelty and Companionship: Conflict in Nineteenth-century Married Life*, London: Routledge.

Herek, G. (1987) 'Can functions be measured? A new perspective on functional approach to attitudes', *Social Psychology Quarterly* 50: 285–303.

hooks, b. (1990) *Yearning: race, gender and cultural politics*, Boston: South End Press.

—— (1991) 'Marginality as a site of resistance', in Russell Ferguson, Martha Gever, Trinh T. Minh-ha and Cornel West (eds) *Out There: Marginalization and Contemporary Cultures*, pp. 341–4, Cambridge, MA: MIT Press.

Howell, Philip (1993) 'Public space and the public sphere: political theory and the historical geography of modernity', *Environment and Planning D: Society and Space* 11: 303–22.

Lloyd, G. (1984) *The Man of Reason: 'Male' and 'Female' in Western Philosophy*, London: Methuen.

MacKinnon, C. (1989) *Toward a Feminist Theory of the State*, Cambridge, MA: Harvard University Press.

Massey, D. (1993) 'Power geometry and a progressive sense of place', in J. Bird,

B. Curtis, G. Robertson and L. Tickner (eds) *Mapping the Futures: Local Cultures, Global Change*, London: Routledge.

Massumi, Brian (1992) *A User's Guide to Capitalism and Schizophrenia*, Cambridge, MA: MIT Press.

Matthews, G. (1992) *The Rise of Public Woman: Woman's Power and Woman's Place in the United States: 1638–1978*, New York: Oxford University Press.

Mitchell, D. (1995) 'The end of public space? People's Park, definitions of the public, and democracy', *Annals of the Association of American Geographers* 85(1): 108–33.

Munt, S. (1995) 'The lesbian flâneur', in D. Bell and G. Valentine (eds) *Mapping Desire*, London: Routledge.

O'Donovan, K. (1993) 'Sexual divisions in the law', in S. Jackson (ed.) *Women's Studies: A Reader*, New York: Harvester Press.

Pain, R. (1991) 'Space, sexual violence and social control: integrating geographical and feminist analyses of women's fear of crime', *Progress in Human Geography* 15(4): 415–31.

Pence, E. and Shepard, M. (1988) 'Integrating feminist theory and practice: the challenge of the battered women's movement', in K. Yllo and M. Bograd (eds) *Feminist Perspectives on Wife Abuse*, pp. 282–98, Newbury Park, CA: Sage.

Pleck, E. (1987) *Domestic Tyranny: The Making of American Social Policy from Colonial Times Until the Present*, Oxford: Oxford University Press.

Robbins, Bruce (ed.) (1993) *The Phantom Public Sphere*, Minneapolis: University of Minnesota Press.

Saland, S. (New York State Senator) (1994) 'News release', distributed 2 February.

Schneider, E. (1994) 'The violence of privacy', in M. Fineman and R. Mykitiuk (eds) *The Public Nature of Private Violence*, New York: Routledge.

Soja, E. (1989) *Postmodern Geographies*, New York: Verso.

Spivak, G.C. (1988) *In Other Worlds: Essays in Cultural Politics*, New York: Routledge.

Squires, J. (1994) 'Private lives, secluded places: privacy as political possibility', *Environment and Planning D: Society and Space* 12: 387–401.

Stone, L. (1977) *Family, Sex and Marriage in England 1500–1800*, New York: Harper & Row.

Thomas, D. and Beasley, M. (1993) 'Domestic violence as a human rights issue' 15 *Human Rights Quarterly* 36.

Tosh, John (1994) 'What should historians do with masculinity? Reflections on nineteenth-century Britain', *History Workshop Journal* 38 (Autumn).

United States Commission on Civil Rights (1982) *Under the Rule of Thumb*, Washington, DC.

Valentine, G. (1989) 'The geography of women's fear', *Area* 21: 385–90.

—— (1992) '(Hetero)sexing space: lesbian perceptions and experiences of everyday spaces', *Environment and Planning D: Society and Space* 11: 395–413.

—— (1994) 'Girls, girls, girls – contested femininities in everyday spaces', paper presented at the Association of American Geographers Annual Meeting, San Francisco.

Young, I.M. (1990) *Justice and the Politics of Difference*, Princeton, NJ: Princeton University Press.

Zorza, J. (1992) 'The criminal law of misdemeanor domestic violence, 1970–1990', 83 *Journal of Criminal Law and Criminology* 83–46.

CHAPTER 9 (RE)NEGOTIATING THE 'HETEROSEXUAL STREET'

Aurand, S., Addessa, R. and Bush, C. (1985) *Violence and Discrimination against Philadelphia Lesbian and Gay People: A Study by the Philadelphia Lesbian and Gay Task Force*, Philadelphia: Philadelphia Lesbian and Gay Task Force.

Bell, D. (1995) 'Perverse dynamics, sexual citizenship and the transformation of intimacy', in D. Bell and G. Valentine (eds) *Mapping Desire: Geographies of Sexualities*, London: Routledge.

Bell, D. and Valentine, G. (1995) 'The sexed self: strategies of performance, sites of resistance', in S. Pile and N. Thrift (eds) *Mapping the Subject: Geographies of Cultural Transformation*, London: Routledge.

Bell, D., Binnie, J., Cream, J. and Valentine, G. (1994) 'All hyped up and no place to go', *Gender, Place and Culture* 1: 37–47.

Berrill, K. (1992) 'Anti-gay violence and victimisation in the United States: an overview', in G. Herek and K. Berrill (eds) *Hate Crimes: Confronting Violence against Lesbians and Gay Men*, London: Sage.

Bradby, B. (1993) 'Lesbians and popular music: does it matter who is singing?' in G. Griffin (ed.) *Outwrite: Lesbianism and Popular Culture*, London: Pluto Press.

Bristow, J. (1989) 'Being gay: politics, pleasure, identity', *New Formations* 9: 61–81.

Bunch, C. (1991) 'Not for lesbians only', in S. Gunew (ed.) *A Reader in Feminist Knowledge*, London: Routledge.

Butler, J. (1990) *Gender Trouble: Feminism and the Subversion of Identity*, London: Routledge.

Comstock, G.D. (1989) 'Victims of anti-gay/lesbian violence', *Journal of Interpersonal Violence* 4: 101–6.

—— (1991) *Violence Against Lesbians and Gay Men*, New York: Colombia University Press.

Coward, R. (1984) *Female Desire: Women's Sexuality Today*, London: Paladin Books.

Davis, T. (1995) 'The diversity of Queer Politics and the redefinition of sexual identity and community in urban spaces', in D. Bell and G. Valentine (eds) *Mapping Desire: Geographies of Sexualities*, London: Routledge.

Diva (1994) 'Lesbian avengers campaign kicks off with series of zaps', 9 October.

Evans, D. (1993) *Sexual Citizenship: The Material Constructions of Sexualities*, London: Routledge.

Foley, C. (1994) *Sexuality and the State: Human Rights Violations against Lesbians, Gays, Bisexuals and Transgendered People*, London: National Council for Civil Liberties.

Forrest, K V. (1984) *Amateur City*, London: Pandora.

Foucault, M. (1979) *Discipline and Punish*, New York: Vintage.

Hayes, J (1976) 'Gayspeak', *Quarterly Journal of Speech* 62: 256–66.

Herek, G (1988) 'Heterosexuals' attitudes toward lesbians and gay men: correlates and gender differences', *Journal of Sex Research* 25(4): 451–77.

Hopkins, R. (1994) 'Sisterly subversion', *Rouge* 18: 22–3.

London Lesbian Offensive Group (1984) 'Anti-lesbianism in the Women's Liberation Movement' in H. Kanter, S. Lefanu, S. Shah and C. Spedding (eds) *Sweeping Statements: Writings from the Women's Liberation Movement 1981–83*, London: The Women's Press.

Massey, D. (1991) 'The political place of locality studies', *Environment & Planning A* 23: 267–81.

McDowell, L. (1995) 'Bodywork: heterosexual gender performances in city work-places', in D. Bell and G. Valentine (eds) *Mapping Desire: Geographies of Sexualities*, London: Routledge.

Meono-Picado, P. (1995) 'Redefining the barricades: Latina lesbian politics and the appropriation of public space', paper presented at the New Horizons in Feminist Geography conference, Kentucky, USA. Available from author at the School of Geography, Clark University, USA.

Mitchell, D. (1995) 'The end of public space? People's Park, definitions of the public and democracy', *Annals of the Association of American Geographers* 85(1): 108–33.

Munt, S. (1995) 'The lesbian flâneur', in D. Bell and G. Valentine (eds) *Mapping Desire: Geographies of Sexualities*, London: Routledge.

Painter, D.S. (1981) 'Recognition among lesbians in straight settings', in J.W. Chesbro (ed.) *Gayspeak*, New York: Pilgrim Press.

Probyn, E. (1992) 'Technologising the self: a future anterior for cultural studies', in

L. Grossberg, C. Nelson and P. Treichler (eds) *Cultural Studies*, pp. 501–11, London: Routledge.

—— (1995) 'Lesbians in space. Gender, sex and the structure of missing', *Gender, Place and Culture* 2(1): 77–84.

Reekie, G. (1993) *Temptations: Sex, Selling and the Department Store*, St Leonards, NSW, Australia: Allen & Unwin.

Rose, G. (1993) *Feminism and Geography: The Limits of Geographical Knowledge*, Cambridge: Polity Press.

Rubin, G. (1984) 'Thinking sex: notes towards a radical theory of the politics of sexuality', in C. Vance (ed.) *Pleasure and Danger: Exploring Female Sexuality*, London: Routledge.

Scene Out (1991) 'It 'Asda' be perv-ect . . . ', December, issue 33: 9.

Sinfield, A. (1993) 'Should there be lesbian and gay intellectuals?' in J. Bristow and A. Wilson (eds) *Activating Theory: Lesbian, Gay and Bisexual Politics*, London: Lawrence & Wishart.

Smyth, C. (1992) *Lesbians Talk Queer Notions*, London: Scarlett Press.

Valentine, G. (1993) '(Hetero)sexing space: lesbian perceptions and experience of everyday spaces', *Environment and Planning D: Society and Space* 11: 395–413.

—— (1995) 'Creating transgressive space: the music of kd lang', *Transactions of the Institute of British Geography* 20: 474–85.

Walker, L. (1995) 'More than just skin-deep: fem(me)ininity and the subversion of identity', *Gender, Place and Culture* 2(1):71–6.

Ware, V. (1992) *Beyond the Pale: White Women, Racism and History*, London: Verso.

Weeks, J. (1992) 'Changing sexual and personal values in the age of AIDS', paper presented at the Forum on Sexuality Conference, Sexual Cultures in Europe, Amsterdam, June.

CHAPTER 10 RENEGOTIATING THE SOCIAL/SEXUAL IDENTITIES OF PLACES

Anderson, Craig L. (1982) 'Males as sexual assault victims: multiple levels of trauma', *Journal of Homosexuality* 7: 145–62.

Berrill, K.T. (1992) 'Anti-gay violence and victimization in the United States: an overview', in Gregory M. Herek and Kevin T. Berrill (eds) *Hate Crimes: Confronting Violence Against Lesbians and Gay Men*, pp. 19–46, Newbury Park, CA: Sage.

Bursik, Robert J., Jr. and Grasmick, Harold G. (1991) *Neighborhoods and Crime: The Dimensions of Effective Community Control*, New York: Lexington Books.

Comstock, Gary David (1991) *Violence Against Lesbians and Gay Men*, New York: Columbia University Press.

D'Emilio, J. (1981) 'Gay politics, gay communities: the San Francisco experience', *Socialist Review* 55: 77–104.

Davis, T. (1994) 'Gay territories, queer spaces: reinforcing and destroying the power of the "gay ghetto"', paper presented at Syracuse University Symposia, Syracuse, NY, 11 February.

England, Kim (1991) 'Gender relations and the spatial structure of the city', *Geoforum* 22(2): 135–47.

Herek, G. and Berrill, K.T. (1992) 'Primary and secondary victimization in anti-gay hate crimes: official response and public policy', in Gregory M. Herek and Kevin T. Berrill (eds) *Hate Crimes: Confronting Violence Against Lesbians and Gay Men*, pp. 289–305, Newbury Park, CA: Sage.

McDowell, L. (1983) 'Towards an understanding of the gender division of urban space', *Environment and Planning D: Society and Space* 1: 59–72.

National Gay and Lesbian Task Force (1994) *Anti-Gay/Lesbian Violence, Victimization, and Defamation in 1993*, Washington: National Gay and Lesbian Task Force Policy Institute.

Pain, R. (1991) 'Space, sexual violence and social control: integrating geographical and feminist analyses of women's fear of crime', *Progress in Human Geography* 15: 415–32.

Peake, L. (1993) '"Race" and sexuality: challenging the patriarchal structuring of urban social space', *Environment and Planning D: Society and Space* 11: 415–32.

Riger, Stephanie (1991) 'On Women', in Dan A. Lewis (ed.) *Reactions to Crime*, Newbury Park, CA: Sage.

Smith, S. (1986) *Crime, Space and Society*, Cambridge: Cambridge University Press.

Valentine, G. (1989) 'The geography of women's fear', *Area* 21: 385–90.

—— (1993) '(Hetero)sexing space: lesbian perceptions and experiences of everyday spaces', *Environment and Planning D: Society and Space* 11: 395–413.

van der Wurff, Adri, van Staalduinen, Leendert and Stringer, Peter (1989) 'Fear of crime in residential environments: testing a social psychological model', *Journal of Social Psychology* 129(2): 141–60.

CHAPTER 11 ON BEING NOT EVEN ANYWHERE NEAR 'THE PROJECT'

Adler, S. and Brenner, J. (1992) 'Gender and space: lesbians and gay men in the city', *International Journal of Urban and Regional Research* 16: 24–34.

Anderson, P. (1980) *Debates in Western Marxism*, London: Verso.

Barrett, M. and Phillips, A. (eds) (1992) *Destabilizing Theory: Contemporary Feminist Debates*, Stanford, CA: Stanford University Press.

Bart, P.B. and Moran, E.G. (1993) *Violence Against Women: The Bloody Footprints*, London: Sage.

Bechdel, Alison (1994) *Dykes to Watch Out For: 1994 Calendar*, Ithaca, NY: Firebrand Books.

Bell, D. (1991) 'Insignificant others: lesbian and gay geographies', *Area* 23: 323–9.

Bell, D., Binnie, J., Cream, J. and Valentine, G. (1994) 'All hyped up and no place to go', *Gender, Place and Culture* 1: 31–48.

Bondi, L. (1992) 'Politics of identity', paper presented at the Annual Meeting of the Institute of British Geographers, Swansea, Wales, January.

Bowlby, S., Lewis, J., McDowell, L. and Foord, J. (1989) 'The geography of gender', in R. Peet and N. Thrift (eds) *New Models in Geography*, pp. 157–75, Boston: Unwin Hyman.

Chouinard, V. (1989) 'Class formation, conflict and housing policies', *International Journal of Urban and Regional Research* 13(2): 390–416.

—— (1992) 'Geography and law: which ways ahead?' paper presented in special session on Geography and Law, Institute of British Geographers Annual Meeting, Swansea, Wales, January.

—— (1994) 'Geography, law and legal struggles: which ways ahead?' *Progress in Human Geography* 18: 415–40.

Christopherson, S. (1989) 'On being outside "the project"', *Antipode* 21: 83–89.

Dear, M. (1981) 'Social and spatial reproduction of the mentally ill', in M. Dear and A.J. Scott (eds) *Urbanization and Planning in Capitalist Societies*, pp. 481–97, London: Methuen.

—— (1988) 'The postmodern challenge: reconstructing human geography', *Transactions of the Institute of British Geographers* 13: 262–74.

Dear, M. and Wolch, J. (1987) *Landscapes of Despair: From Deinstitutionalization to Homelessness*, Princeton, NJ: Princeton University Press.

Deutsche, R. (1991) 'Boys' town', *Environment and Planning D: Society and Space* 9: 5–30.

Disabled Persons for Employment Equity (1992) unpublished flyer, Toronto, Canada.

Dobash, R.E. and Dobash, R.P. (1992) *Women, Violence and Social Change*, New York: Routledge.

Dorn, M. (1994) 'Disability as spatial dissidence: a cultural geography of the stigmatized body', unpublished MA thesis, Department of Geography, Pennsylvania State University.

Duncan, S. and Goodwin, M. (1988) *The Local State and Uneven Development: Behind the Local Government Crisis*, Cambridge: Polity Press.

Duncan, S. and Ley, D. (1982) 'Structural Marxism and human geography', *Annals of the Association of American Geographers* 72: 30–59.

Elliott, S. (1992) 'Psychosocial impacts in populations exposed to solid waste facilities', unpublished PhD dissertation, Geography Department, McMaster University, Hamilton, Canada.

England, K.V.L. (1994) 'Getting personal: reflexivity, positionality, and feminist research', *Professional Geographer* 46: 80–9.

Faludi, S. (1992) *Backlash: The Undeclared War on American Women*, New York: Crown.

Fincher, R. (1991) 'Caring for workers' dependents: gender, class and local state practice in Melbourne', *Political Geography Quarterly* 10: 356–81.

Findlay, S. and Randall, M. (eds) (1988) 'Feminist perspectives on the Canadian state', Special Issue on Women and the State, *Resources for Feminist Research*, 17(3).

Fraser, N. (1989) *Unruly Practices: Power, Discourse and Gender in Contemporary Social Thought*, Minneapolis: University of Minnesota Press.

Gilbert, M. (1994) 'The politics of location: doing feminist research at "home"', *The Professional Geographer* 46: 90–6.

Golledge, R. (1993) 'Geography and the disabled: a survey with special reference to vision impaired and blind populations', *Transactions of the Institute of British Geographers* 18: 63–85.

Hahn, H. (1986) 'Disability and the urban environment: a perspective on Los Angeles', *Environment and Planning D: Society and Space* 4: 273–88.

—— (1989) 'Disability and the reproduction of bodily images: the dynamics of human appearances', in J. Wolch and M. Dear (eds) *The Power of Geography: How Territory Shapes Social Life*, pp. 370–88, Boston: Unwin Hyman.

Harvey D. (1989) *The Condition of Postmodernity*, Oxford and Cambridge, MA: Blackwell.

—— (1992) 'Postmodern morality plays', *Antipode* 24: 300–26.

hooks, b. (1990) *Black Looks: Race and Representation*, Toronto: Between the Lines.

Johnston, G. (1979) *Which Way Out of the Men's Room?* South Brunswick, NJ: A.S. Barnes & Co. Inc.

Johnston, J. (1973) *Lesbian Nation: The Feminist Solution*, New York: Simon & Schuster.

Katz, C. (1994) 'Playing the field: questions of fieldwork in geography', *The Professional Geographer* 46: 67–72.

Knopp, L. (1987) 'Social theory, social movements and public policy: recent accomplishments of the gay and lesbian movements in Minneapolis, Minnesota', *International Journal of Urban and Regional Research* 11: 243–61.

—— (1990a) 'Social consequences of homosexuality', *Geographical Magazine* 62: 20–5.

—— (1990b) 'Some theoretical implications of gay involvement in an urban land market', *Political Geography Quarterly* 9: 337–52.

Knox, P.L. (1991) 'The restless urban landscape: economic and sociocultural change and the transformation of Metropolitan Washington, D.C.', *Annals of the American Association of Geographers* 81: 101–209.

Kobayashi, A. (1994) 'Coloring the field: gender, "race", and the politics of fieldwork', *The Professional Geographer* 46: 73–80.

Kobayashi, A. and Mackenzie, S. (eds) (1989) *Remaking Human Geography*, Boston: Unwin Hyman.

Lauria, M. and Knopp, L. (1985) 'Toward and analysis of the role of gay communities in the urban renaissance', *Urban Geography* 6: 152–69.

Ley, D. and Mills, C. (1993) 'Can there be a postmodernism of resistance in the urban landscape?' in P.L. Knox (ed.) *The Restless Urban Landscape*, pp. 255–78, Englewood Cliffs, NJ: Prentice-Hall.

Massey, D. (1991) 'Flexible sexism', *Environment and Planning D: Society and Space* 9: 31–57.

Masuda, S. and Ridington, J. (1992) *Meeting Our Needs: An Access Manual for Transition Houses*, Toronto: DAWN Canada.

McDowell, L. (1992a) 'Doing gender: feminism, feminists, and research methods in human geography', *Transactions of the Institute of British Geographers* 17: 399–416.

—— (1992b) 'Multiple voices: speaking from inside and outside "The Project"', *Antipode* 24: 56–72.

—— (1993a) 'Space, place and gender relations, Part I: Feminist empiricism and the geography of social relations', *Progress in Human Geography* 17: 157–79.

—— (1993b) 'Space, place and gender relations, Part II: Identity difference, feminist geometries and geographies', *Progress in Human Geography* 17: 305–18.

Miliband, R. (1991) *Divided Societies: Class Struggle in Contemporary Capitalism*, Oxford and New York: Oxford University Press.

Morris, J. (1991) *Pride Against Prejudice*, London: The Women's Press.

Mouffe, C. (1988) 'Radical democracy: modern or postmodern?' in A. Ross (ed.) *Universal Abandon? the Politics of Postmodernism*, pp. 31–45, Minneapolis: University of Minnesota Press.

Murgatroyd, L., Savage, M., Shapiro, D., Urry, J., Walby, S., Warde, A. and Mark-Lawson, J. (1985) *Localities, Class and Gender*, London: Pion Ltd.

Nast, H.J. (1994) 'Women in the field: critical feminist methodologies and theoretical perspectives', *The Professional Geographer* 46: 54–66.

National Council of Welfare (1990) *Women and Poverty Revisited*, Ottawa: Minister of Supply and Services.

Oliver, M. (1990) *The Politics of Disablement*, New York: Macmillan.

Palmer, B. (1990) *Descent into Discourse*, Philadelphia: Temple University Press.

Peake, L. (1993) '"Race" and sexuality: challenging the patriarchal structure of urban social space', *Environment and Planning D: Society and Space* 11: 415–32.

Peet, R. and Thrift, N. (eds) (1989a) *New Models in Geography*, Vols 1 and 2, Boston: Unwin Hyman.

—— (1989b) 'Political economy and human geography', in R. Peet and N. Thrift (eds) *New Models in Geography*, pp. 3–29, Boston: Unwin Hyman.

Pratt, G. (1993) 'Feminist geography', *Progress Report in Urban Geography* 13: 385–91.

Pratt, G. and Hanson, S. (1994) 'Geography and the construction of difference', *Gender, Place and Culture: a journal of feminist geography* 1(1): 5–30.

Rooney, F. and Israel, P. (eds) (1985) 'Women and disability', Special Issue of *Resources for Feminist Research* 4(1).

Ross, A. (ed.) (1988) *Universal Abandon? The Politics of Postmodernism*, Minneapolis: University of Minnesota Press.

Ross, B. (1990) 'The house that Jill built: lesbian feminists organizing in Toronto, 1976–1980', *Feminist Review* 35: 75–91.

Staeheli, L.A. and Lawson V.A. (1994) 'A discussion of "women in the field": the politics of feminist fieldwork', *The Professional Geographer* 46: 96–102.

Statistics Canada (1993) *The Violence Against Women Survey: Highlights*, Ottawa: Minister Responsible for Statistics Canada.

Taylor, S.M. (1989) 'Community exclusion of the mentally ill', in J. Wolch and M. Dear (eds) *The Power of Geography: How Territory Shapes Social Life*, pp. 316–30, Boston: Unwin Hyman.

Thompson, E.P. (1978) 'The poverty of theory or an orrery of errors', in *The Poverty of Theory and Other Essays*, pp. 1–210, New York: Monthly Review Press.

Valentine, G. (1992) 'Towards a geography of the lesbian community', paper presented at the Women in Cities conference, Hamburg, 10 April.

—— (1993a) '"Desperately seeking Susan": a geography of lesbian friendships', *Area* 25: 109–16.

——(1993b) 'Negotiating and managing multiple sexual identities: lesbian time–space strategies', *Transactions of the Institute of British Geographers* 18: 237–48.

—— (1993c) '(Hetero)sexing space: lesbian perceptions and experiences of everyday spaces', *Society and Space* 11: 395–413.

Walker, G. (1990) *Family Violence and the Women's Movement: The Conceptual Politics of Struggle*, Toronto: University of Toronto Press.

Winchester, H.P.M. and White, P.E. (1988) 'The location of marginalized groups in the inner city', *Environment and Planning D: Society and Space* 6: 37–54.

Wolch, J. and Dear, M. (eds) (1989) *The Power of Geography: How Territory Shapes Social Life*, Boston: Unwin Hyman.

CHAPTER 12 ENGENDERING RACE RESEARCH

Adilman, T. (1984) 'A preliminary sketch of Chinese women and work in British Columbia 1858–1950', in B. Latham, R. Latham and R. Pazdro (eds) *Not Just Pin Money*, pp. 53–78, Victoria: Camusun College.

Amos, V. and Parmar, P. (1984) 'Challenging imperial feminisms', *Feminist Review* 32.

Anderson, K. (1990) 'Chinatown re-oriented: a critical analysis of recent redevelopment schemes in a Sydney and Melbourne enclave', *Australian Geographical Studies* 28(2): 137–54.

——(1991) *Vancouver's Chinatown: Racial Discourse in Canada, 1875–1980*, Montreal: McGill-Queens University Press.

Anthias, F. and Yuval-Davis, N. (1992) *Race, Nation, Gender, Colour and Class and the Anti-racist Struggle*, London: Routledge.

Back, L. (1994) 'The "white negro" revisited: race and masculinities in South London', in A. Cornwall and N. Lindisfarne (eds) *Dislocating Masculinities*, pp. 172–83, London: Routledge.

Bear, L. (1994) 'Miscegenations of modernity: constructing European respectability and race in the Indian railway colony, 1857–1931', *Women's History Review* 3(4): 531–48.

Bhabha, H. (1990a) 'The other question: difference, discrimination and the discourse of colonialism', in R. Ferguson, M. Gever, T. Minh-ha and C. West (eds) *Out There: Marginalisation and Contemporary Cultures*, pp. 71–88, New York: New Museum of Contemporary Art and Massachusetts Institute of Technology.

—— (ed.) (1990b) *Nation and Narration*, London: Routledge.

Bottomley, G. (1991) 'Representing the "second generation": subjects, objects and ways of knowing', in G. Bottomley, M. de Lepervanche and J. Martin (eds) *Intersexions: Gender/Class/Culture/Ethnicity*, pp. 92–110, Sydney: Allen & Unwin.

Bottomley, G., and de Lepervanche, M. (1984) *Ethnicity, Class, and Gender in Australia*, Sydney: Allen & Unwin.

Bottomley, G., de Lepervanche, M. and Martin, J. (eds) (1991) *Intersexions: Gender/Class/Culture/Ethnicity*, Sydney: Allen & Unwin.

Butler, J. (1990) *Gender Trouble: Feminism and the Subversion of Identity*, New York: Routledge.

Canada (1885a) *Debates of the House of Commons*.

——(1885b) *Sessional Papers*, No. 54a, Royal Commission on Chinese Immigration.

——(1887) *Debates of the House of Commons*, 31 May.

——(1922) *Debates of the House of Commons*, 8 May.

Chong, D. (1994) *The Concubine's Children: Portrait of a Family Divided*, Harmondsworth: Penguin.

Clifford, J. (1988) *The Predicament of Culture*, Cambridge, MA: Harvard University Press.

Cohen, S. (1972) *Folk Devils and Moral Panics*, London: MacGibbon & Kee.

Collins, P. (1991) 'Learning from the outsider within: the sociological significance of black feminist thought', in J. Hartman and E. Messer-Davidow (eds) *(En)gendering*

Knowledge: Feminists in the Academe, pp. 40–65, Knoxville: University of Tennessee Press.

Donald, J. and Rattansi, A. (eds) (1992) *'Race', culture and difference*, Milton Keynes and London: The Open University/Sage.

Donaldson, L. (1992) *Decolonizing Feminisms: Race, Gender and Empire-building*, Chapel Hill and London: The University of North Carolina Press.

Foucault, M. (1972) *The Archaeology of Knowledge*, transl. by A.M. Sheridan Smith, New York: Harper & Row.

—— (1979) *Discipline and Punish*, New York: Viking.

hooks, b. (1981) *Ain't I a Woman? Black Women and Politics*, London: South End Press.

—— (1991) *Yearning: Race, Gender and Cultural Politics*, London: Turnaround.

Jackson, P. (1994) 'Black male: advertising and the cultural politics of masculinity', *Gender, Place and Culture* 1(1): 49–60.

Jennett, C. and Randal, S. (eds) (1987) *Three Worlds of Inequality: Race, Class and Gender*, South Melbourne: Macmillan.

Kobayashi, A. and Peake, L. (1994) 'Unnatural discourse: "race" and gender in geography', *Gender, Place and Culture* 1(2): 225–43.

Kwong, P. (forthcoming) 'Back to basics: politics of organizing Chinese women garment workers', *Social Policy*.

Larbalestier, J. (1991) 'Through their own eyes: an interpretation of Aboriginal women's writing', in G. Bottomley, M. de Lepervanche and J. Martin (eds) *Intersexions: Gender/Class/Culture/Ethnicity*, pp. 75–91, Sydney: Allen & Unwin.

Lee, S. (1990) *Disappearing Moon Cafe*, Vancouver: Douglas & McIntyre.

Li, P. (1988) *The Chinese in Canada*, Toronto: Oxford University Press.

Lowe, J. (1991) *Critical Terrains: French and British Orientalisms*, Ithaca and London: Cornell University Press.

Nicholson, L. (ed.) (1990) *Feminism/Postmodernism*, New York: Routledge.

Pateman, C. (1988) *The Sexual Contract*, Cambridge: Polity Press.

Pettman, J. (1992) *Living in the Margins: Racism, Sexism and Feminism in Australia*, Sydney: Allen & Unwin.

Plumwood, V. (1993) *Feminism and the Mastery of Nature*, London: Routledge.

Poovey, M. *Uneven Developments: the Ideological Work of Gender in Modern Victorian England*, Chicago: Chicago University Press, 1989.

Pratt, G. (1993) 'Reflections on feminist empirics', *Antipode* 25(1): 51–63.

Rose, G. (1993) *Feminism and Geography: The Limits of Geographical Knowledge*, Cambridge: Polity Press.

Sacks, K. (1989) 'Towards a unified theory of class, race and gender', *American Ethnologist* 16(3): 534–50.

Said, E. (1978) *Orientalism*, New York: Random House.

Satzewich, V. (1989) 'Racisms: the reactions to Chinese migrants in Canada at the turn of the century', *International Sociology* 4(3): 311–27.

Schaffer, K. (1995) 'The Elisa Fraser story and constructions of gender, race and class in Australian culture', *Hecate* 17(1): 136–49.

Singleton, C. (1989) 'Race and gender in feminist theory', *Sage* 6(1): 12–17.

Spelman, E. (1978) *Inessential Women: Problems of Exclusion in Feminist Thought*, Boston: Beacon Press, 1988.

Stepan, N. (1991) *The Hour of Eugenics: Race, Gender and Nation in Latin America*, Ithaca and London: Cornell University Press.

Thomas, N. (1994) *Colonialism's Culture: Anthropology, Travel and Government*, Melbourne: Melbourne University Press.

Walby, S. (1992) 'Post-post modernism? Theorizing social complexity', in M. Barrett and A. Phillips (eds) *Destabilizing Theory: Contemporary Feminist Debates*, pp. 31–52 Cambridge: Polity Press.

Ware, V. (1992) *Beyond the Pale: White Women, Racism and History*, London: Verso.

Ward, P. (1978) *White Canada Forever: Popular Attitudes and Public Policy Toward Orientals in British Columbia*, Montreal: McGill-Queens University Press.

Wickberg, E., Con, H., Johnson, G. and Willmott, W. (1982) *From China to Canada: A History of the Chinese Communities in Canada*, Toronto: McClelland & Stewart.

Yee, P. (1988) *Saltwater City*, Vancouver: Douglas & McIntyre.

Yuval-Davis, N. (1991) 'The citizenship debate: women, the state and ethnic processes', *Feminist Review* 39: 58–68.

CHAPTER 13 DISPLACING THE FIELD IN FIELDWORK

Bell, David, Binnie, Jon, Cream, Julia and Valentine, Gill (1994) 'All hyped up and no place to go', *Gender, Place and Culture* 1(1): 31–47.

Butler, Judith (1994) 'Gender as performance', *Radical Philosophy* 67: 32–9.

Clifford, James (1988) *The Predicament of Culture: Twentieth-century Ethnography*, Cambridge, MA: Harvard University Press.

—— (1990) 'Notes on (field)notes', in Roger Sanjek (ed.) *The Makings of Anthropology*, Ithaca, NY: Cornell University Press.

Cosgrove, Denis (1993) 'Commentary', *Annals of the Association of American Geographers* 83(3): 515–16.

Cosgrove, Denis and Daniels, Stephen (1989) 'Fieldwork as theatre: a week's performance in Venice and its region', *Journal of Geography in Higher Education* 13(2): 169–82.

de Lauretis, Teresa (1987) *Technologies of Gender: Essays on Theory, Film and Fiction*, Bloomington: Indiana University Press.

Derrida, Jacques (1981) *Dissemination*, transl. Barbara Johnson, Chicago: Chicago University Press.

—— (1982) 'White mythology: metaphor in the text of philosophy', in *Margins of Philosophy*, pp. 109–36, transl. Alan Bass, Chicago: University of Chicago Press.

Dyck, Isabel (1993) 'Ethnography: a feminist method?' *The Canadian Geographer* 37(1): 52–7.

Emberly, Julia V. (1993) *Thresholds of Difference: Feminist Critique, Native Women's Writings, Postcolonial Theory*, Toronto: University of Toronto Press.

England, Kim V.L. (1994) 'Getting personal: reflexivity, positionality and feminist research', *The Professional Geographer* 46(1): 80–9.

Foucault, Michel (1980) 'Questions on geography', in Colin Gordon (ed.) *Power/Knowledge: Selected Interviews and Other Writings, 1972–1977*, pp. 63–77, New York: Pantheon.

Freud, Sigmund (1959) *Beyond the Pleasure Principle*, transl. James Strachey, New York: Bantam. (Orig. 1928.)

Geertz, Clifford (1973) *The Interpretation of Cultures*, New York: Basic Books.

—— (1988) *Works and Lives: The Anthropologist as Author*, Stanford, CA: Stanford University Press.

Gilbert, Melissa R. (1994) 'The politics of location: doing feminist research at "home"', *The Professional Geographer* 46(1): 90–6.

Gordon, Deborah (1988) 'Writing culture, writing feminism: the poetics and politics of experimental ethnography', *Inscriptions* 3/4: 7–24.

Gregory, Derek (1994) *Geographical Imaginations*, New York: Blackwell.

Grosz, Elizabeth (1990) 'Inscriptions and body-maps: representations and the corporeal', in Terry Threadgold and Anne Cranny-Francis (eds) *Feminine/Masculine and Representation*, pp. 62–74, Sydney: Allen & Unwin.

Haraway, Donna (1989) *Primate Visions: Gender, Race and Nature in the World of Modern Science*, New York: Routledge.

—— (1991a) 'Situated knowledges: the science question in feminism and the privilege of partial perspective', in *Simians, Cyborgs and Women: The Reinvention of Nature*, pp. 183–202, New York: Routledge.

—— (1991b) 'The contest for primate nature: daughters of man-the-hunter in the field, 1960–80', in *Simians, Cyborgs and Women: The Reinvention of Nature,* pp. 149–81, New York: Routledge.

—— (1991c) 'A cyborg manifesto: science, technology, and socialist-feminism in the late twentieth century', in *Simians, Cyborgs and Women: The Reinvention of Nature,* pp. 149–81, New York: Routledge.

Harding, Sandra (1991) 'Who knows? Identities and feminist epistemology', in Joan Hartman and Ellen Messer-Davidous (eds) *(En)Gendering Knowledge,* pp. 110–15, Knoxville: University of Tennessee Press.

hooks, bell (1990) *Yearning: race, gender and cultural politics,* Boston: South End Press.

Hyndman, Jennifer (1995) 'Solo feminist geography: a lesson in space', *Antipode* 22(2) 197–208.

Jarosz, Lucy (1992) 'Constructing the Dark Continent: metaphor as a geographic representation of Africa', *Geografiska Annaler* 74, B (2): 102–15.

Jay, Martin (1990) 'Fieldwork and theorizing in intellectual history', *Theory and Society* 19: 311–21.

Katz, Cindi (1992) 'All the world is staged: intellectuals and the process of ethnography', *Environment and Planning D: Society and Space,* 10: 495–510.

—— (1994) 'Playing the field: questions of fieldwork in geography', *The Professional Geographer* 46(1): 66–72.

Katz, Cindi and Smith, Neil (1993) 'Grounding metaphor: towards a spatialized politics', in Michael Keith and Steve Pile (eds) *Place and the Politics of Identity,* pp. 67–83, New York: Routledge.

Kirby, Kathleen M. (1993) 'Thinking through the boundary: the politics of location, subjects and space', *boundary 2* 20(2): 173–89.

Kobayashi, Audrey (1994) 'Coloring the field: gender, "race", and the politics of field-work', *The Professional Geographer* 46(1): 73–80.

LaCapra, Dominick (1980) 'Who rules metaphor?' *diacritics* 10(4): 15–28.

Lefebvre, Henri (1991) *The Production of Space,* transl. Donald Nicholson-Smith, Oxford: Blackwell.

Madge, Clare (1993) 'Boundary disputes: comments on Sidaway (1992)', *Area* 25(3): 294–9.

Marcus, George (1992) '"More (critically) reflexive than thou": the current identity politics of representation', *Environment and Planning D: Society and Space* 10: 489–93.

Matless, David (1989) 'Appendix: deconstructive commentary', *Journal of Geography in Higher Education* 13(2): 182.

McDowell, Linda (1992) 'Multiple voices: speaking from inside and outside "The Project"', *Antipode* 24(1): 56–72.

—— (1993) 'Space, place and gender relations: Part 1. Feminist empiricism and the geography of social relations', *Progress in Human Geography* 17(2): 157–79.

Mitchell, Timothy (1988) *Colonising Egypt,* Cambridge: Cambridge University Press.

—— (1990) 'Everyday metaphors of power', *Theory and Society* 19: 545–77.

Moss, Pamela (1993) 'Feminism as method', *The Canadian Geographer* 37(1): 48–9.

—— (1994) 'Spatially differentiated conceptions of gender in the workplace', *Studies in Political Economy* 43: 79–116.

Nast, Heidi J. (1994) 'Opening remarks on "Women in the field"', *The Professional Geographer* 46(1): 54–66.

Nast, Heidi J. and Blum, Virginia (under review) 'This space which is not one: Henri Lefebvre and Jacques Lacan', *Environment and Planning D: Society and Space.*

Pred, Allan (1984) 'From here and now to there and then: some notes on diffusions, defusions and disillusions', in M.D. Billinge, D.J. Gregory and R.L. Martin (eds) *Recollections of a Revolution: Geography as Spatial Science,* pp. 86–103, London: Macmillan.

Price, Marie and Lewis, Martin (1993) 'The reinvention of cultural geography', *Annals of the Association of Human Geographers* 83(1): 1–18.

Pringle, Rosemary (1988) *Secretaries Talk: Sexuality, Power and Work*, London: Verso.

Probyn, Elspeth (1993) *Sexing the Self: Gendered Positions in Cultural Studies*, New York: Routledge.

Rabinow, Paul (1977) *Reflections on Fieldwork in Morocco*, Berkeley: University of California Press.

Ricouer, Paul (1979) *The Rule of Metaphor: Multidisciplinary Studies in the Creation of Meaning in Language*, transl. Robert Czerny with Kathleen McLaughlin and John Costello, Toronto: Toronto University Press.

Rose, Damaris (1993) 'On feminism, method and methods in human geography: an idiosyncratic overview', *The Canadian Geographer* 37(1) 57–61.

Rose, Gillian (1993) *Feminism and Geography: The Limits of Geographical Knowledge*, Cambridge: Polity Press.

Rubin, Barbara (1975) 'Prostitution in Nevada', *Annals of the Association of American Geographers* 65: 113–15.

Sauer, C.O. (1956) 'The education of a geographer', *Annals of the American Association of Geographers* 46(3): 287–99.

Scott, Joan W. (1992) 'Experience', in Judith Butler and Joan W. Scott, *Feminists Theorize the Political*, New York: Routledge.

Skeels, Anna Clare (1993) 'A passage to premodernity: Carl Sauer repositioned in the field', unpublished Masters thesis, University of British Columbia, Vancouver, Canada.

Sparke, Matthew (1994a) 'Writing on patriarchal missiles: the chauvinism of the Gulf War and the limits of critique', *Environment and Planning A* 26: 1061–89.

—— (1994b) 'White mythologies and anemic geographies', *Environment and Planning D: Society and Space* 12: 105–23.

—— (1994c) 'A prism for contemporary capitalism: temporary work as displaced labour as value', *Antipode* 26(4): 295–322.

—— (forthcoming) 'Flexibility for who? Gender and resistance in the traffic of temporary workers', *Gender, Place and Culture*.

Spivak, Gayatri Chakravarty (1976) 'Translator's preface', in Jacques Derrida, *Of Grammatology*, Baltimore: Johns Hopkins University Press.

—— (1992) *Thinking Academic Freedom in Gendered Post-coloniality*, Cape Town: University of Cape Town Press.

Staeheli, Lynn A. and Lawson, Victoria A. (1994) 'A discussion of "women in the field": the politics of feminist fieldwork', *The Professional Geographer* 46(1): 96–102.

Starn, Orin (1994) 'Rethinking the politics of anthropology: the case of the Andes', *Current Anthropology* 35(1): 13–38.

Stewart, Lynn (1995) 'Bodies, visions and spatial politics: a review essay on Henri Lefebvre's *The Production of Space*', *Environment and Planning D: Society and Space* 13(5): 609–15

Stoddart, David R. (1986) *On Geography and its History*, Oxford: Blackwell.

Symanski, Richard (1974) 'Prostitution in Nevada', *Annals of the Association of American Geographers* 64: 357–77.

Visweswaran, Kamala (1994) *Fictions of Feminist Ethnography*, Minneapolis: University of Minnesota Press.

CHAPTER 14 REFLECTIONS ON POSTMODERN FEMINIST SOCIAL RESEARCH

Barrett, M. (1991) *The Politics of Truth: From Marx to Foucault*, Cambridge: Polity Press.

—— (1992) 'Words and things: materialism and method in contemporary feminist analysis', in M. Barrett and A. Phillips (eds) *Destabilizing Theory: Contemporary Feminist Debates*, Stanford, CA: Stanford University Press.

Brown, W. (1991) 'Feminist hesitations, postmodern exposures', *Differences* 3(1): 63–84.

Butler, J. (1990) *Gender Trouble: Feminism and the Subversion of Identity*, New York: Routledge.

D'Amico-Samuels, D. (1991) 'Undoing fieldwork: personal, political, theoretical and methodological implications', in F.Y. Harrison (ed.) *Decolonizing Anthropology: Moving Further Toward an Anthropology for Liberation*, Washington, DC: Association of Black Anthropologists and American Anthropological Association.

Dyck, I. (1993) 'Ethnography: a feminist method?' *The Canadian Geographer* 37(1): 52–7.

Ferguson, K. (1993) *The Man Question: Visions of Subjectivity in Feminist Theory*, Berkeley: University of California Press.

Gibson, K. (1991a) 'Hewers of cake and drawers of tea: women and restructuring on the coalfields of the Bowen Basin', *Economic and Regional Restructuring Research Unit Working Paper 2*, Department of Geography, University of Sydney, NSW 2006 Australia.

—— (1991b) 'Company towns and class processes: a study of Queensland's new coalfields', *Environment and Planning D: Society and Space* 9(3): 285–308.

—— (1992a) 'Hewers of cake and drawers of tea: women, industrial restructuring and class processes on the Central Queensland coalfields', *Rethinking Marxism* 5(4): 29–56.

—— (1992b) '"There is no normal weekend anymore . . ." ' A report to the coal mining communities of Central Queensland on family life and continuous shift work', unpublished report available from author at Dept of Geography and Environmental Science, Monash University, Clayton, Victoria 3168, Australia.

—— (1993) *Different Merry-Go-Rounds: Families, Communities and the 7-day Roster*, Brisbane: Queensland Colliery Employees Union.

Gibson-Graham, J.K. (1994a) '"Stuffed if I know!" Reflections on post-modern feminist social research', *Gender, Place and Culture* 1(2): 205–24.

—— (1994b) 'Beyond capitalism and patriarchy: reflections on political subjectivity', in B. Caine and R. Pringle (eds) *Transitions: New Australian Feminisms*, Sydney: Allen & Unwin.

Grosz, E. (1984) 'Interview with Gayatri Spivak', *Thesis Eleven* 10/11: 175–87.

Harding, S. (1986) *The Science Question in Feminism*, Ithaca: Cornell University Press.

Kondo, D.K. (1990) *Crafting Selves: Power, Gender, and Discourses of Identity in a Japanese Workplace*, Chicago: University of Chicago Press.

McDowell, L. (1992) 'Doing gender: feminism, feminists and research methods in human geography', *Transactions of the Institute of British Geographers* 17: 399–416.

Metcalfe, A. (1987) 'Manning the mines: organizing women out of class struggle', *Australian Feminist Studies* 4: 73–96.

Nast, H., Katz, C., Kobayashi, A., England, K.V.L., Gilbert, M., Staeheli, L.A. and Lawson, V.A. (1994) 'Women in the field: critical feminist methodologies and theoretical perspectives', *The Professional Geographer* 46(1): 54–102.

Pratt, G. (1993) 'Reflections on poststructuralism and feminist empirics, theory and practice', *Antipode* 25(1): 51–63.

Reinharz, S. (1992) *Feminist Methods in Social Research*, New York: Oxford University Press.

Rose, D. (1993) 'On feminism, method and methods in human geography: an idiosyncratic overview', *The Canadian Geographer* 37(1): 57–61.

Spivak, G. (1989) 'A response to *The Difference Within: Feminism and Critical Theory*,' in E. Meese and A. Parker (eds) *The Difference Within: Feminism and Critical Theory*, Amsterdam: John Benjamin.

Sturmey, R. (1989) *Women and Services in Remote Company Dominated Mining Towns*, Armidale: The Rural Development Centre, University of New England, NSW 2351.

Weedon, C. (1987) *Feminist Practice and Poststructuralist Theory*, Oxford: Blackwell.

Young, I.M. (1990) 'The ideal of community and the politics of difference', in L. Nicholson (ed.) *Feminism/Postmodernism*, New York: Routledge.

CONCLUSION

Butler, Judith (1992) 'Contingent foundations: feminism and the question of "Postmodernism"', in Judith Butler and Joan Scott (eds) *Feminists Theorize the Political*, pp. 22–40, New York: Routledge.

Davis, Tim (1995) 'The diversity of queer politics and the redefinition of sexual identity and community in urban spaces' in David Bell and Gill Valentine (eds) *Mapping Desire*, pp. 284–303, London: Routledge.

Massumi, Brian (1992) *A User's Guide to Capitalism and Schizophrenia*, Cambridge, MA: MIT Press.

Phillips, Anne (1993) *Democracy and Difference*, Oxford: Polity Press.

INDEX